11-28-73

THE ARCTIC HIGHWAY

THE ARCTIC HIGHWAY

A road and its setting

John Douglas

DAVID & CHARLES : NEWTON ABBOT
STACKPOLE BOOKS : HARRISBURG

ISBN 0 7153 5269 5 (*Great Britain*)
ISBN 0 8117 0172 7 (*United States*)

This edition first published in 1972
in Great Britain by David & Charles (Publishers) Limited
Newton Abbot Devon
in the United States by Stackpole Books Harrisburg Pa

Printed in Great Britain by
W. J. Holman Limited Dawlish

1784373

To my parents

Contents

TABLES

Illustrations

PLATES

Plates 33 (above), 87, 121 and 140 are reproduced by permission of A. S. Norsk Telegrambyrå. All other photographs are by the author.

MAPS

Introduction

A little over a hundred years ago the potential traveller to North Norway would probably have been somewhat deterred had he consulted Price's *A Road Book for Tourists*: 'This route (Trondhjeim to Hammerfest) follows the road to the Nausen as far as Hun, and has only one stage by land beyond that place. Such a journey for nearly 900 miles performed in open boats, in all weather, through a most desolate and sterile region, has little to compensate the fatigue and expense of the undertaking.'

Even forty years ago the same journey would have combined road and sea travel and was unlikely to be tackled by any but the most adventurous. Today, in summer, one can catch the North Norway Bus in Fauske and be carried in comfort along the *Arctic Highway* to arrive four days later in Kirkenes on the Soviet border: a journey through a landscape of magnificent varied scenery and human interest along a road which is unparalleled by any other of the world's great highways. No other continuous highway so penetrates the arctic and reaches so near to the Pole. No other road so well demonstrates man's determination and ability to conquer a polar clime.

The road begins as the European Route 6 (E6) in Malmö, Sweden. It follows the coast to Oslo, then strikes across country through the delightful Gudbrandsdal to Trondheim. From here it winds its way through Nord-Trondelag and Sør-Nordland to reach Mo i Rana and the Polar Circle. It is the route from Mo to Kirkenes with which this book is concerned: nine hundred miles of *Arctic Highway* in *Nord Norge* (comprising the *fylker*, or counties, of Nordland, Troms and Finnmark).

There is nowhere else that is quite like North Norway—

11

certainly nowhere else within the polar circles. To appreciate the individuality of the *Arctic Highway* is impossible unless the distinctiveness of the geographical background, physical and human, is understood. The first chapter of this book provides the background. Following an account of the history of the *Arctic Highway* in Chapter 2, the greater part of the book describes the whole length of the road from Mo i Rana to Kirkenes, and all the Highway's major branches as well as some of the more alluring minor roads. A short chapter describes the impact which the Highway's construction has had on those most interesting of people, the Lapps, and finally there is a look into the future and a reference to tourism.

For those interested in statistics, a number of tables are included in the Appendices. The maps are intended only to assist in the location of the more important places along and just off the Highway: to follow the text with greater interest and understanding, the reader is recommended to use one of the many large scale maps of North Norway such as the excellent Cappelen 1:400,000 series.

THE ARCTIC HIGHWAY

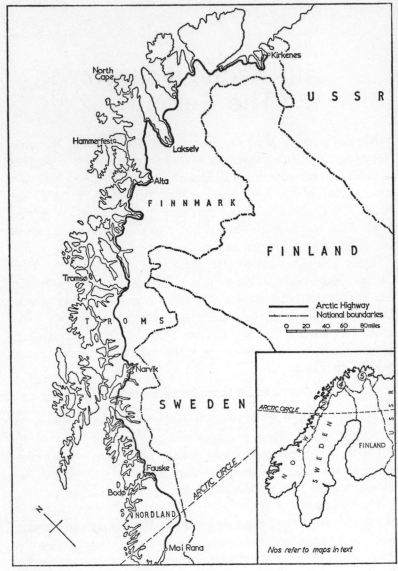

Arctic Highway
National boundaries

0   20   40   60   80 miles

Nos refer to maps in text

MAP 1

Chapter One

The Setting

Norway's Arctic Highway, between Mo i Rana and Kirkenes (see Map 1), threads its way across rather more than four degrees of polar latitude covering a distance of over 900 miles. All but a short section of the road, just north of Mo i Rana, is inside the Arctic Circle and at its most northerly point it comes to within $19\frac{1}{2}°$ of the North Pole. The Highway goes as far north as central Greenland and as far east as Istanbul.

There is no other comparable highway inside either of the Polar Circles. Greenland and Antarctica are virtually roadless, while in Alaska and northern Canada few roads penetrate the far north. Even the recent extensions to the famous Alaskan Highway only brush latitude 66° North. In the USSR the arctic roads are generally subordinate to air or river routes and rarely reach beyond the Circle.

The Physical Background

The Arctic Highway owes its existence to the unique character of North Norway. Here is an environment that has many of the attributes of the true polar world, yet also has a climate and topography which, since the retreat of the Quarternary ice sheets, has allowed man to live so remarkably close to the North Pole.

The singularity of the climate is of prime importance. North Norway's winter temperatures are some 20°C higher in January than the average for the latitude. Narvik's lowest mean monthly temperature is only a few degrees below zero, yet comparable stations in Canada's Far North experience −30°C and the notorious

Verkhoyansk, on a similar latitude in the USSR, has temperatures below −13°C for fully seven months of the year. Although Kirkenes, at the north-eastern end of the Highway, has a February mean of −10°C, generally one has to go inland or climb into the mountains to find really arctic temperatures. Thus, away from the Highway on the Finnmark plateau, mean minima of −14.5°C occur in the Lapp settlements of Karasjok and Kautokeino and, elsewhere, shelter from the sea causes local pockets of coldness.

The basic cause of North Norway's *relatively* mild winter conditions is the combined effect of a warm current, the North Atlantic Drift, and the incidence of westerly airstreams over most of the region. This is not the place to discuss the complexities of these climatic factors, but two points deserve a mention. The effect of the North Atlantic Drift is only partly to bring warm water to Norway's shores. Perhaps its greatest contribution is to separate the cold polar waters from the warm sector of the eastern Atlantic, which in turn increases the temperature of the westerly winds. The result is the longest ice-free coast in the polar world. Secondly, on those atypical occasions when the airflow is not from the west, temperatures plummet to true arctic depths. It is little comfort, as one freezes in the biting frost of a *jernnatt* (iron night), to be told about *average* temperatures.

Springtime can be very cold but very beautiful. Much of the sun's radiant energy is used to melt the accumulated snows of winter but, as the rivers swell and life returns, it is not difficult to appreciate the special significance that this season has for all who live in the North. The days lengthen and the sun rises higher in the sky : spring suddenly becomes summer.

In June, everywhere along the Highway where the road touches the coast, the mean temperature exceeds 8°C and by July—almost always the warmest month—the temperature has risen above 12°C to as high as 13.6°C at Alta. As in all seasons, airstreams from the north or east are especially influential in providing unusually cold or warm conditions. It is common to experience really high temperatures in summer with a light easterly airstream. (I have recorded 27°C as early as 10 am just east of Alta in August.)

Autumn, with its marvellous colourful beauty, arrives early in September, and by October mean monthly temperatures are scarcely above zero even at the coast. Tana, on the Finnmark section of the Highway, is fully 5°C below freezing by November and only Narvik, of the towns along the road, has to wait until mid-November before daily means below freezing are the rule.

Precipitation in North Norway is much less than is popularly supposed; indeed, parts are distinctly dry. The coastlands have an annual average of about 40in but totals inland are very much lower. Much of the snow and rainfall comes with depressions which cross the Norwegian Sea to impinge on the high west coast. Their greatest frequency is autumn and early winter, when heavy snow results. However, because of the alignment of the coast, in the northern fylke of Finnmark and even in north Troms, the regime is reversed and summer is the wettest season. One of the wettest points along the Highway is Mo i Rana with over 50in per annum but, away from the exposed west coast, Alta receives only just over 15in. On the southward sloping *vidda* of Finnmark the totals are as low as 11 to 13in, the sort of figures one associates with Alice Springs in the Australian Desert. Nevertheless, the precipitation is especially effective and accumulations of snow bring severe problems.

One inch of rain is the approximate equivalent of 12in of freshly fallen snow and along the Highway it is not unknown to experience snowfalls even in August on the higher plateau sections. Overall, between 40 and 50 per cent of the precipitation occurs as snow which lies for over half the year at sea level in the northern two fylker but for only four to five months in Nordland. Depths of lying snow vary according to drift-producing winds but three feet (without drifting) is common in late winter.

Two unattractive features of the climate are the wind chill experienced in polar airstreams and the general cloudiness of North Norway. Cloud base levels are usually higher in summer than in winter but clouds can descend to 400ft over the Finnmark vidda in July, adding to the hazards of flying or even driving over plateau roads.

B

Some reference must be made to that peculiarly arctic pheno-
menon: the midnight sun. If conditions allow (and it must be
admitted that rarely do they), the sun can be seen as a complete
disc at the North Cape from 14 May to 30 July and even at
Narvik for as long a period as 26 May to 19 July. The low but
little changing angle of the sun gives exceptionally small diurnal
ranges of temperature. In summer the air is already warm when
most people are rising and it is pleasant to walk or read by sun-
light late into the evening. But in winter there are the long hours
of depressing darkness when artificial light is necessary throughout the day.

The topography of North Norway is less kind to the inhabitants than is the climate. Consisting of two distinct regions,
the older Baltic Shield of south and east Finnmark and the
Caledonian mountain system of the west coast, arctic Norway is
something of a challenge to the geologist. The area's potential
as a source of mineral ores has yet to be fully explored and exploited but it is the mineral wealth that has often exerted and will
continue to exercise a powerful influence on communications.

The retreat of the ice sheets is *recent* in geological terms,
occurring perhaps some 8,000 to 10,000 years ago and leaving
an indelible mark on the landscape. The pre-glacial river valleys
were over-deepened and straightened into giant U-shaped
troughs which form the deep fjords of the west coast and the
wider, softer inlets of the north. As the ice fronts retreated and
the glaciers shortened, deposits of sands, gravels and clays were
laid down together with erratic boulders, often choking the valley
mouths.

With the weight of ice removed from the land surface, changes in sea level occurred. As the land rose, so too did the sea and,
while a balance was maintained, strand lines were cut in the
fjord sides. These now frequently form long narrow shelves along
the coast, most suitable for communications and settlement.

From even the simplest atlas map, North Norway's present
physiography can be seen as mountainous with an indented
coast. Off-shore lie countless *skerries* or islands, some rising

steeply from the sea, others no more than low *strand-flat* platforms with scarcely any soil cover. Between the islands and the mainland is the *skjærgård*, the waters so well known to Norwegian fishermen. The fractured, sinuous coast is immensely long, offering only rare sites for villages or courses for roads.

Nordland and Troms are squeezed into a narrow coastal belt by the high ranges of the Scandinavian Uplands. Rivers flow swiftly over falls and rapids towards the sea and deeply incise their valleys. The plateau areas are often remote and the broadest, Saltfjellet, is further isolated by one of those relics of the Ice Age, the huge Svartisen ice-cap.

Finnmark is unlike its neighbouring fylker. It is broader and less rugged, consisting largely of a vast plateau or *vidde* which slopes gently southward from a high coast broken not by narrow inlets but by wide fjords. The scenery is like none other in Norway, more likely recalling northern Finland or Sweden. It is in Finnmark that one finds the real tundra and the excitement of a polar world.

The Fauna and Flora

In an area so scantily populated as North Norway, the vegetation is largely natural. Regional variations occur with conifers in the southern fylker, being replaced by birch in the north, and altitude as well as latitude is a determining factor. All the same, the variety of vegetation is a local rather than a regional feature. Variations in soil and the ever-changing micro-climates produce a flora which is never wholly the same as one passes from district to district.

Certain limiting factors are common: the very short but vigorous growing season, the general acidity of the soils, the water-logged nature of the ground, the steep soil-free slopes and the wind-swept plateaux. Forests occur where shelter, latitude, altitude and soil depth permit. Nordland and Troms carry large stands of conifers, chiefly pine. This species extends into Finnmark where the northernmost pine forest in the world is found bordering the Highway in Stabbursdalen, but in this fylke the birch predominates among the trees. Even in Troms and Nordland the pine stands are fragmented and their commercial usefulness reduced. The birch varies from the scrub birch of the least favoured positions, through dwarf birch, to taller trees in the meadow woods. Pine and birch combine in the south where transitional zones are found on the mountain sides. Other trees such as alder and types of arctic willow also occur but compete unfavourably with dominant stands.

Peat bogs, with tufted sedge grass and delicate cotton grass, give way to mosses and lichens where mountain vegetation borders the permanent snow line. Much of Finnmark is true tundra and often waterlogged. Where better drained, heath plants prevail and edible berries are common. The delicious orange cloudberries (moltebær), sometimes known as arctic strawberries, are a much sought-after delicacy.

Large areas, because of exposure, altitude or slope, are bare of vegetation and the country rock is open to rapid weathering.

In the animal world of North Norway, a small rodent occupies a place of interest far higher than his size would suggest. I refer, of course, to the lemming. This little member of the *muridae* family, with its black and yellow fur, is rarely seen by anyone but the patient naturalist, for it is nocturnal by habit, small in size and relatively uncommon. But in those seasons when the natural growth rate of the lemming reaches explosive proportions, they seem to dominate the mammal population. Their suicidal marine excursions are well-known. Their over-population problem is solved dramatically and the quiet years return again when there is food for all. Quite why the reproduction rate so increases is not fully understood, but it is noteworthy that other rodents are similarly and simultaneously affected and that a correlation with climatic conditions and, therefore, food supply can be shown. Voles, rats and hares are other common rodents.

Other mammals are somewhat shy. Wolves and wolverines, and even bears, still occupy the mountain areas of these three northern fylker but they are rare today and seen only when exceptionally severe winters curtail their normal supplies of food. Other beasts of prey such as foxes are more commonly seen and they are remarkably fearless even when closely approached. On the whole, however, few wild animals are seen by the traveller or resident in North Norway unless he makes a special effort to seek them out. There is some hunting of elk and roedeer but the reindeer are now herded.

The birds of the area are notable for their variety and their uneven distribution. The super-abundance of sea food, along with a greater cover of vegetation at the coast, gives rise to far larger numbers and greater varieties on the seaward margins of the region. Small willow tits, golden eagles (only recently protected), oyster-catchers and arctic warblers; birds of prey, town birds and sea birds—all have their homes in arctic Norway. Thousands of eider ducks and single sparrow hawks, millions of puffins in what amounts to a single colony; yet inland in the mountains of Troms and Nordland or on the plateau of Finnmark a single bird can be a rare sight.

The insect world is rich. Much of the more tundra-like areas as well as the forests and boglands prove to be ideal breeding grounds for a form of life which can easily withstand the long cold winters. Mosquitoes in their millions, as is so common in all sub-polar regions, are a scourge to both man and beast when temperature and wind conditions encourage them to move freely and greedily. Gadflies, ants, cockroaches and beetles are among the other insects which abound to delight the entomologist and plague the traveller.

From the point of view of North Norway's economy, fish are indisputably the most important form of animal life. The rivers and lakes provide some of the best salmon and trout fishing in the world, where even the least skilled and most impatient angler is assured of a catch. The sea fisheries are world famous and, more often than not, the *raison d'être* of the settlement. The mixture of warm Atlantic and cold polar waters is ideal and produces gigantic shoals of cod and herring, the two most important species, as well as capelin, coalfish and haddock. Whales and seals are also to be found, although the former have decreased greatly in recent decades.

The People and their Economy

North Norway has been spasmodically settled for at least 9,000 years. Stone-age man favoured Finnmark where he could engage in fishing, hunting and trapping, and to which he had easier access over the flatter vidda. The populations of Nordland and Troms have fluctuated more dramatically. Generally speaking, settlements able to engage in the dual-economy of fishing and agriculture survived more easily than those villages reliant upon fishing alone.

Populations have been swelled since early times by people moving into the coastal districts from the interior. Notable among these groups have been the Finns or *Kvæns* in the north, the Skolt Lapps in the north-east and the Swedish Lapps in Nordland. Most of these movements were brought about by political or economic pressures and this is partly true of the migrations from South Norway. Even today, many of the people of the three northern fylker are only first or second generation North Norwegians. However, recent settlers have found more positive attractions in the expanding economy, and their mixed origins enrich the culture of the region.

Two further demographic factors are important. Firstly, the trend of populations to move out of the more remote parts of the north to the larger settlements and, secondly, the general drift of North Norwegians to the South. Indeed, if it were not for the growth in numbers through immigration, North Norway would suffer from serious depopulation.

The distribution of the people is most uneven. Fully ninety per cent live within fifteen minutes' walking distance of the sea, and certain larger towns dominate all other settlement. From Mo i Rana northwards, five towns overshadow all others in Nordland and Troms. These are Mo, Bodø and Narvik in Nordland and, in Troms, the capital Tromsø and Harstad. But in Finnmark at least six towns, Hammerfest, Alta, Vardø, Vadsø, Kirkenes and Honningsvåg, vie with one another on a more or less equal footing. It is not without significance that only four of the eleven towns named above lie on the Highway.

Fishing remains the mainstay of the North's economy, although the proportion employed in this industry continues to decline. Centralisation and increases in scale are today's key-notes of an industry which in the past has played the greatest part in attracting man north of the Arctic Circle. The fish-rich rivers, *skjærgård* and Norwegian Sea draw mariners from all the major West European fishing countries and the sheltered inlets provide a multitude of safe harbours. Today, the greatest threat to the industry lies in the fishing policy of the European Economic Community which Norway has applied to join.

While fishing reigns supreme in the northern economy, other activities necessarily supplement its inadequate income. Many find employment in more than one industry. Fishing and farming; farming and forestry; fishing and building—these are but a few of the possible combinations. In the last quarter of a century there has been a sharp rise in the service industries and State intervention has enabled one or two large-scale industries such as the A/S Norsk Jernverk plant at Mo i Rana to come into being.

Topographically and climatically, farming is handicapped to an almost prohibitive degree. Limited to the coast or off-shore islands and to a few valley lowlands, limited by a growing season which never exceeds five months and often drops below three, farming is restricted spatially and in range of product. Indeed, less than one per cent of North Norway is cultivated in any way. Fodder crops are of the greatest importance, for more than half the agricultural income is derived from cattle or cattle products.

No farming scene in summer is without its racks of drying hay and a cow grazing in a field is a rare sight. Instead, cattle wander through woodland, along roadsides and on upland pastures in summer and are stall-fed in winter. Although most farmers own some woodland, which is cut as the need arises, large-scale lumbering is very restricted.

Mining is an industry which has a very direct influence on

communications in North Norway. Two minerals, iron and copper, dominate. Iron is presently worked near Kirkenes and Mo i Rana, while copper is important in Sulitjelma and west of Kautokeino. Other settlements, such as Kåfjord in Finnmark, have survived the working-out of ores.

Since the early 1950s, State help for the economy of North Norway has been on a grand scale. Direct investment in energy production, schools and communications, coupled with subsidies and a separate fiscal policy for the North, all have helped to strengthen the region's economy and reduce the drift of population to the South.

Communications : the Problems

It is no exaggeration to say that Norway, north of Mo i Rana, is something of a road engineer's or transport economist's nightmare. How else can one view a territory of some 38,585 square miles and 400,000 souls, almost entirely inside the Arctic Circle and with a coastline so indented that it would stretch from Great Britain to South America. This is to say nothing of the fact that two-thirds of the land is above 1,000ft, and that there are over 50,000 islands off the shore, many of which are inhabited.

The communication problems may be divided into those attributable to the harsh physical environment and those resulting from difficulties in the social and human geography of the region. The latter are probably the more important.

Only the briefest mention will be made of communications other than roads but it must be appreciated that all the transportation media are at once complementary and competitive. It would be foolish to assume that roads play as great a part in North Norway as in more temperate and less littoral areas. The sea provides the traditional routes for the transport of people and goods and the rôle of the aeroplane is only slightly less important here than in other arctic regions.

Physical Problems

Strictly, there are two physical handicaps to communications :

the climate and the topography. Generally unhampered by ice and only occasionally interrupted by fog and gales, sea transport has remarkably little to fear from the weather. Airways are only rather less kindly treated for, although some runways may be closed temporarily through snow or fog, winter—with its vast expanses of snow—provides an almost limitless landing ground for aircraft equipped with skis. Turbulence is greatly aggravated by the contiguity of sea, mountains and ice but, although it makes the flying of light aircraft difficult, airliners are little affected. Similarly, it is the smaller craft that are inhibited by winter darkness.

The shelter of the *skjærgård,* the indented coast and the ease with which large ships can move along the coast and into the fjords in all seasons, all favour sea transportation. The German Fleet was quick to take advantage of the natural harbours during the 1940-5 war and there are few places where the small Norwegian fishing vessels cannot readily find shelter.

Topographical handicaps are more noticeable when one turns to air travel. There is little flat ground for airfields and only Bodø, Banak, Bardufoss and Alta have airports within the town limits. Such large towns as Hammerfest, Narvik and Harstad are still without a runway for landplanes. However, the abundance of water surfaces favours the seaplane and, in summer, floats replace skis on light aircraft. Very few of the larger settlements, or even small hamlets in Finnmark, are without their mooring place for seaplanes. Nevertheless, the rugged relief generally makes flying difficult. To see the tiny seaplanes flying in half light through gorge-like valleys or large airliners manœuvre into limited approach paths, is to witness the skill of experts.

Land transport offers the alternative of road or railway. Not surprisingly, the latter more easily contends with the problems of climate. Giant rotary ploughs and wooden tunnels or snow-sheds (*see plate, page 52*) make what railways there are all but immune to the severe winter weather. The mountains have taxed the skill and resourcefulness of the railway engineers but the twenty-six mile line from Narvik up to the Swedish frontier is a magnificent

The Arctic Highway

example of what is possible. By way of tunnels and sharp curves above steep precipices, a rise of over 1,700ft is accomplished. All the same, the uninviting terrain north of Straumen on Sørfolda has restricted railway building north of this fjord to mineral lines justified on special economic grounds.

But what of roads? With the overwhelming proportion of roads in North Norway constructed of gravel, any change in weather adversely affects the surface. Rainfall, snowfall, high temperatures, dry periods, frost—all play a part in increasing the cost and difficulties of construction and maintenance.

Arctic Norway's snowfall is not the greatest problem faced by the road engineer but it is the most persistent. There are three attendant difficulties:

    i. Prevention of unmanageable accumulations of snow on the road surface.

    ii. Removal of such snow as has accumulated.

    iii. Spring melt of snow.

Rarely can the routeing of roads be such that areas of heavy snow are entirely avoided. Instead, choice is often restricted to that of particular valley sides or trans-fjell cols in order to avoid local drifting. Aspect is often critical.

Because avoidance of areas subject to heavy snow or drifting is usually impossible, preventative measures have to be taken. Most effective is the raising of the path on an embankment some five to ten feet above the immediate surroundings. In this way the road does not act as a gully into which snow will blow and the ditches on either side form traps to collect the snow naturally or as it is removed from the road surface by ploughs.

Shelter from snow can be afforded by tunnels and by fences. The former are costly and rare. The German occupying forces in the 1940s experimented unsuccessfully with wooden tunnels but today's constructions are primarily built to forge a passage through the mountains and protection from snowfall is an incidental advantage. Half-tunnels or rock canopies are common where roads are blasted out of the fjord walls but massive wooden snow fences are the most usual expedient.

These open wooden lattices, sometimes over twenty feet in height, are effectively placed to stop snow blowing on to the road surface. The siting of the fences, often in rows, is critical and they are frequently found at wind-swept fjord heads and cols, or on exposed plateaux.

In order to allow continuous use of roadways by conventional vehicles, snow ploughs and blowers must be employed. The cost is great but the method successful, as is shown by the steady decrease in recent years of the sections of Arctic Highway closed in winter. Down-slope ploughing is desirable on slopes greater than about 1 in 15 and, if the gradient exceeds 1 in 10, up-slope ploughing may be impossible. Maintenance stations are, therefore, best located near the summits of hills. Speed is essential in preventing drifting, but should this occur, then hand-digging to weaken the drift is followed by frontal attack.

A foot or so of packed snow is usually left on the road, but where the road is not raised—as is the case with much of the Highway in Finnmark—the accumulation is often much greater until even the wooden marker posts are almost submerged. Snow removed from the road builds up at the edges of these sections and the path is narrowed by giant walls which survive the general spring thaw. Worse still, the roads become gullies and snow traps between the walls.

Despite these various precautions, many of North Norway's roads, including two significant sections of the Arctic Highway, have to be closed in winter for as much as five months, from mid-December to the end of April.

When the snow does melt, the roads are often in their worst state and closure to all vehicles may be necessary or axle loads may have to be reduced. Melt water erodes the surface and carves channels into the softened gravel, while in areas where permafrost underlies the surface, the upper layers become *active* (ie unstable). Fortunately, permafrost is sporadic in North Norway and the problems encountered in other arctic regions are less acute here.

Frost heaving and premature melt of snow followed by

re-freezing are other spring problems, as are snow and rock avalanches. The chance of a vehicle being struck by falling rocks may be small but accidents do occur when drivers run into unexpected obstructions. The fjord roads of Troms and Nordland are some of the most affected by rock falls and the avalanches of the Lyngenfjord section of the Highway are notorious.

Streams swollen by the spring melt of snow add another hazard to roads. The heavily loaded waters are a danger to the countless bridges along all highways and flooding of low-lying sections of roads is common.

Rainfall may act on roads in the same way as melt water or it may be the agent by which natural filtration of the road gravels is initiated. The erosion of road surfaces is notably worse where the construction is of water-bound gravel and the damage is greatest on bends, on shallow slopes, at narrow sections and where heavy vehicles break up the surface. Water tends to collect in the furrows made by wheel tracks and the furrows are deepest where the narrowness of the road restricts their number to two or possibly three. These furrows carry miniature rivers after rain and pot holes are scoured out as in the bed of a stream.

In the construction of gravel roads, the coarseness of material increases with depth, but if that material is porous, then water soaking through tends to carry the lighter gravels downwards, exposing the larger fragments at the top. A reduction in the porosity of the road, by using oil-bound gravels or putting on an asphalt surface, can obviously lessen the filtration and erosive effects of rain. But constructional improvements are costly and gravel surfaces, even if oil-bound, are larkely to be the most common in arctic Norway for many years to come.

Very heavy rain can undermine the foundations of the roads, especially where unconsolidated material is exposed to water running off mountain slopes. Mass wasting of road embankments is not unknown and has to be guarded against by frequent inspection and repair.

Dry weather in summer causes a different sort of deterioration of gravel roads. The light surface material is subject to deflation.

Dry gravel, losing its coherence, is easily blown off the surface, leaving the coarse material behind. In these conditions, vehicles become agents of transportation and erosion and commonly will leave a cloud of dust stretching over a hundred yards in their wake.

Vehicles such as the *snowmobile* or *hoverhawk* can cope with the winter conditions of arctic Norway without the need for conventional roads, but the general situation is not suited to the construction of a network of snow roads such as the *avtozimniki* of Siberia. Only where there is a complementary river transport system in summer are such roads viable.

The topographical problems for roads are much the same as those for railways, reduced only by the former's ability to tackle superior gradients. The greatest problem of all is undoubtedly the length and character of the coast. Not only is it deeply indented but there are considerable stretches where there is not even a hint of coastal plain, shelf or *strand-flat*. The mountain drops straight into the sea.

Half Norway's longest fjords (exceeding forty miles) fracture the coast north of Mo i Rana. Porsangen and Varangerfjorden, for example, are both over sixty miles long. Off-shore islands offer a similar challenge to land communications. Two settlements only a few miles apart may have to be linked by a long road winding its way round a fjord head or finding a passage through an intervening mountain range (see Appendix C Table IX).

Ferries and bridges are only a partial solution. Ferries are the traditional means of linking roads on either side of a fjord or connecting routes between mainland and island. However, they tend to delay traffic and restrict the use of roads and are, therefore, being replaced by bridges where this is possible (see, for examples, Chapter 8 Table II).

The construction of tunnels under the fjords or straits has, up to now, been far too expensive, but forward planning has included the costing of such a tunnel for Tromsø. Quite apart from cost, the difficulties in the way of tunnels of this kind are chiefly those of physical geography. In almost every case where a tunnel would be helpful, it would be impracticable because of the near impossibility of constructing an approach road in mountainous country.

The mountains themselves form a real barrier to roads. The Scandinavian Uplands separating Sweden from Norway are only breached by two roads north of Mo i Rana at the present time. One road leads into Sweden from Mo, the other into the Finnish panhandle from Skibotn. In Finnmark, where the southerly sloping vidda is less troublesome, again only two roads run into Finland. These roads are supplemented by numerous footpaths and tracks but these are incapable of carrying motor traffic. Such mountains as there are within Norwegian territory have to be circumscribed by the road network.

Although altitudes reached by main roads are generally lower in the north than in South Norway, gradients are comparable. The Arctic Highway has slopes up to 1 in 7 although, in theory, there is a maximum limit at 1 in 12. The irregular gradients of many river valleys used by roads often means that the natural slope cannot be used. Instead the road must be constructed to bypass the breaks in slope.

To reduce gradients a winding path is often necessary, but where the vidder are crossed or in some of the broader valleys of Finnmark, such as the Tana, opportunities are offered and seized for Roman-like directness. Straightening has been a feature of post-war development and usually the old road is left almost intact, forming a not always accessible lay-by. Curvature reduction is often combined with the construction of newly aligned bridges which are widened to new, higher standards (*see plate, page 51*).

Socio-economic Problems

The human geography and social structure of North Norway present similar problems to all the transportation media but the solutions each has to offer are often very different. Communications by sea, even with the introduction of the Express Steamers (*Hurtigruten*) at the end of the last century, are unavoidably slow and require, for efficiency, a largely coastal economy and settlement or an integrated system of land transport. In fact, the greatest drawback to sea transport is that it encourages the *status quo* in North Norway and inhibits the development of the interior. The traditional way of *colonizing* and servicing the north, this form of transport is rapidly becoming less and less economically viable except for long-haul bulk cargo, generally arising from mineral workings. Thus the ore ships to and from Mo, Narvik and Kirkenes contrast in their profitability with small intra-coastal vessels and regular services for general cargo and passengers. Only by a large State subsidy has this form of transport survived so long with an apparent but unreal efficiency.

Airways, too, are not without their problems although their rise—from nothing—in the post-war period has been remarkable. Despite an annual passenger growth rate of around 15 per cent, the services are generally uneconomic, especially on minor routes. The airfields, limited in number and capacity, the intra-national character of the services and the lack of an integrated local network, all are handicaps. Scheduled local services, such as that inaugurated by Widerøe's in 1968 in the Mo i Rana–Bodø area, are steps in the right direction but there is still much to be done. Helpful to air transport is the dual military–civil function of some landing fields but the use as emergency runways of strengthened and widened sections of road, such as one finds in northern Sweden and Finland, is unknown in arctic Norway.

Light aircraft could well play an even larger part in the transport system of the north. At present their use is largely restricted to emergencies (such as the carriage of the sick and injured to hospital) and to pleasure flying, often associated with angling.

Arctic railways were a nineteenth-century dream and a twentieth-century reality in Norway. All the same, the total length of line constructed is very small (scarcely 200 miles, north of Mo i Rana). Only the *Nordland* (Trondheim to Bodø) railway is part of the national system. The other lines (*Ofotbanen*: to Narvik from Sweden, *Sulitjelmbanen*: Sulitjelma to Finneid, and the *A/S Sydvaranger*: Bjørnevatn to Kirkenes) were all, in origin, mineral ore railways although the line into Narvik carries a substantial number of passengers and is linked to the Swedish national system.

When the *Nordland* railway reached Bodø in 1962 an upsurge in passenger traffic was expected. This has failed to materialise and, mainly because of competition from road transport, the real value of this line is for long-haul freight and passenger traffic, although in this respect sea transport is the real competitor. Local rail traffic is unprofitable and some railway stations built along the line to Bodø never even opened.

There is little doubt that roads offer the most comprehensive, flexible and adaptable network of communications in North Norway. The days when the population was content with the natural *roads* of the sea and river are gone for ever. Without a road system the interior must remain a neglected hinterland and the development and sustentation of the north is in jeopardy.

The non-physical difficulties of roads are chiefly those of finance. In a territory so vast as arctic Norway, with a population density of about ten persons per square mile, the road costs per mile per inhabitant are staggering. The tax paying capacity of the populace is small as is their political influence. Oslo is nearer the Po Plain of Italy than it is to the north of the Finnmark vidda. Little wonder that North Norwegians sometimes feel like neglected colonials: a feeling which, it must be said, is largely unjustified.

The need for an improved road system stems from the necessity or desirability to:

i. Link the scattered communities such that the spheres of influence of the towns may be enlarged, their services extended, and a fuller life enjoyed. This is especially important in an area where depopulation is a constant threat.

ii. Supplement or replace water transport, already running at a loss, and allow the limited rail and air services to be fully functional.

iii. Speed the supply of goods by providing a rational low-cost means of freight carriage.

Page 33 (*above*) A Highway bus. Early summer and the snow still lies at the Highway's edge in Finnmark. The rack on the front of the bus is often used to carry bicycles. (*below*) Repairs to the Highway on Sennalandet. The Highway's surface is still water-bound gravel and can rapidly deteriorate.

Page 34 (*above*) Trensgel Bridge and the Highway skirting Sørfolda. A complex series of bridges and tunnels replaced the old Røsvik-Bonnåsjøen ferry in 1966. (*below*) View out of Kalvik tunnel—the longest of the series of tunnels.

iv. Cause some redistribution of settlement away from the relatively congested areas of the coast towards the interior. The under-development of Finnmark is a case in point.

v. Diversify the economy by providing an infra-structure for developing industries. Tourism is an obvious beneficiary of an improved road system but so too are manufacturing industries and even farming.

vi. Assist in the administration and defence of the region. The military significance of the north is arguable but Norway's £140m defence budget in 1968-9 included an extra £14m to improve the NE border defences with Russia.

But for all that, to establish the need is not to pay the cost. Only such towns as Bodø, Narvik and Mo i Rana can make substantial contributions through kommune funds. The remaining cost must be met by the State.

Expenditure in recent years has increased at a slower rate than traffic and only slightly more steeply than costs. Road maintenance frequently takes as large a share of available resources as do improvements and new constructions (see Appendix A Table III). Per capita expenditure tends to be least in Troms (density of population: thirteen per square mile) and greatest in Finnmark (density four per square mile). The great extent of the northernmost fylke compared with Troms' rather more close knit communities along the coast help to account for the difference. Area for area, however, Finnmark is the poor relation in terms of road building. Comparative figures for expenditure on the Arctic Highway are given for a representative year in Appendix A Table IV. However, just as the human and physical geographies of the three fylker are not entirely comparable, neither are their highway needs. The significance of the Arctic Highway to Finnmark is lessened by its real need for interior development (see Chapter 8).

Other figures for expenditure on the Arctic Highway are shown in Table V (Appendix A) but it must be emphasised again that much of the money goes on maintenance. The cost of *standing still* is exacerbated by the restricted period of the year

c

during which repairs can be effected and by the nature of the surface of the roads. But, because an oil-bound gravel road costs about 30,000 Nkr per kilometre and a macadamed surface infinitely more, a compromise has to be reached between the ideal surface and the demand for further mileage. Nevertheless, as surface improvements are made, maintenance costs should decrease.

When conditions in arctic Norway returned to normal after the last war and the settlements of the north were rebuilt and inhabited again, the traffic using the roads increased out of all proportion to pre-war figures. Admittedly, the Arctic Highway was now more or less complete but, more than that, the post-war population started to turn its back on the sea, perhaps for the first time, and look to land communications to link them with one another and with the rest of Norway and to bring them their share in nationally rising standards of living.

Increases in traffic were most steep in the 1950s and, more especially, in the 1960s despite the fact that car ownership in North Norway remains substantially below the national figure. From the mid-1930s to 1950, bus passenger traffic (measured as kilometre per passenger) increased sevenfold and goods carried by the bus services showed a fourfold increase. The 1950s saw a further doubling of passengers and trebling of goods. Although the increases in the last decade have not been so dramatic, the totals continue to rise so that many millions are now carried by bus. The annual total number of passengers carried along the Highway by the North Norway Bus is currently around 300,000. Traffic census details are given in Appendix B Tables VI to VIII.

Ferries have already been mentioned. They present a special economic problem when they prevent twenty-four hour travel. The total length of ferry connections in the three fylker in 1967 was: Finnmark—18.4km; Troms—68.1km; all Nordland—386.1km. These ferries carried respectively: 2,832,395; 1,588,059 and 338,926 passengers. While ferry statistics only partly reflect road figures, it is worth noting that the number of all vehicles using ferries shows annual increases of about 15 per cent and the increases in freight trucks is often much greater.

Two other socio-economic problems have to be faced: the problem of seasonal variation in traffic flow and the problem of markedly varying traffic densities with exceptional increases towards the urban agglomerations.

Seasonal variations are not as great as might at first be imagined. The reasons seem to be threefold. The reduction of periods of winter road closure and the fact that the North Norwegian, with practised disregard for the elements, will use his car if at all possible, explain the high winter figures. In summer, the increase in tourist traffic is largely offset by the efflux of local inhabitants taking their vacations in warmer climes. The latter point was borne out by a detailed census of traffic carried out in the Mo i Rana district in 1968.

Where the weather seriously hampers traffic, the seasonal variation is clearer. Thus at Skaidi, where the Arctic Highway meets the road to Hammerfest, the average number of vehicles per day for the whole of 1965 was 410, but for the summer months the daily average was 715. Elsewhere the variation is of the order of a 25 to 40 per cent summer increase.

This summer increase in traffic comes at a time when all road maintenance must be carried out. Work is impossible in winter and spring so that 100 per cent of all maintenance and construction has to be fitted into perhaps 30 to 40 per cent of the year. To drive more than 20 miles and not see a road gang at work is something worthy of comment in arctic Norway in the summer months (*see plate, page 33*).

Coping with local variations in traffic densities is an even greater problem. The major traffic census of 1965 confirmed the view that comparative concentrations build up towards the urban centres but that, away from the towns and larger villages, high densities are the exception. In fact, in the mid-1960s, only sixteen extra-urban sections of main highway exceeded 500 vehicles per day (see Appendix B Table VIII). In contrast, many roads have less than 200vpd and even sections of the Arctic Highway carry less than 100 vehicles. The densities vary inversely as the distance from the nearest urban centre; where towns are especially isolated the concentrations are more marked.

North Nordland may be taken as fairly typical of this feature. Thus Mo i Rana, 50 miles from Mosjøen and 153 miles from Bodø, has heavy traffic only within a radius of 15 miles of the town. Similarly, Narvik shows an even greater concentration than Mo. Bodø, on the other hand, has a high density of traffic along the whole of Route 80 to Fauske, 40 miles away. Certain sections of road where one might have expected a high density are in fact little used. The Arctic Highway between Skarberget and Sætran, only 37 miles from Narvik, had a density of less than 100 vehicles per day in 1965, but then, of course, to reach Narvik involved the use of two ferries. Likewise, that rare occurrence, a road to Sweden the road from Mo i Rana to Umbukta and, ultimately, to Tärnaby carries very light traffic.

In Troms, the roads in the Kvængsfjell region are litle used as are those on the western side of Lyngenfjorden. Even the road from Skibotn into the north-west Finnish panhandle has a low density, particularly if one excludes military vehicles. In contrast, the heavily settled Bardudalen and lowlands around Bardufoss have the highest traffic densities in the fylke, unequalled even by the well-used road between Vollan and the capital, Tromsø, or by the roads into Harstad.

Little traffic uses the roads in Finnmark. Apart from the concentrations around the Altafjord, the head of Porsangen, near the capital, Vadsø, and in the neighbourhood of Kirkenes, the roads are often comparatively free of traffic. Of course, a relatively large number of Finnmark's roads are closed in winter but, even in summer the rural roads are remarkably empty. In the mid-1950s in high summer only 100 cars per day were passing through Karasjok on Route 96. Yet at that time there was no alternative road into Finland such as the Kautokeino road, Route 93, of to-day. Even in the late 1960s and now, the situation is much the same.

However, if a rural area is well settled the traffic density increases significantly over quite long distances. The road from

Tanabru to Vadsø, some forty-one miles, carries quite heavy traffic most of the year and there are similar instances in all three fylker.

These varying densities, never high but occasionally very low indeed, present the transport planners with a dilemma. Naturally the roads with the heaviest traffic must be given the better surfaces, the higher standard widths and the greatest attention to maintenance. But this policy leaves little money for the improvement or maintenance of the highways between the larger settlements and serving scattered rural populations. In this way the real problem of providing a *network* of roads remains unsolved, rather it is aggravated. Of course, the rural roads are not entirely neglected. Witness the improvements to Route 93 into Finland, to the North Cape road, to the Arctic Highway east of Skaidi and countless examples throughout the region. Nevertheless, it is the whole system of roads that needs attention, not just relatively short sections.

Investment in the highway system is an investment in the whole of North Norway's future, for it is in this way that the region will be transformed and be able to play its full part in the country's economy. It can fairly be argued that too little attention has been paid to road development in arctic Norway; that too much money has been spent on perpetuating an anachronistic population and transport structure based on coastal settlements and an economy derived from the sea. Traffic densities outside the towns will remain low just as long as the rural roads are poor. The stage has been reached when roads should be improved to encourage higher volumes of traffic rather than wait for more vehicles to prove the need.

Chapter Two

History and Development of the Arctic Highway

Where did the Arctic Highway start? When did the construction of the Highway begin? There can be no definitive answers to these questions. The Highway's history can be pieced together only with difficulty, and even in the last fifty years there are many uncertainties.

As a near continuous road between Mo i Rana and Kirkenes, the Arctic Highway dates from the early 1940s and its planning, in detail, started only some twenty years earlier. Before that, all that existed were sections of road serving isolated communities. Isolated, that is, except by sea communications. The centuries of development of North Norway, from the early settlers to modern times, has been based on a coastal population served by the sea. Not only did the early communities have no roads but they had no need for a road link throughout the province. The economy, the distances and the physical handicaps combined to produce a situation in which a road such as the Arctic Highway had no place.

It must not be thought that the people of North Norway, with feelings of deprivation, waited for advances in engineering and the release of capital for the construction of a road link with the south. This is far from the truth. Significantly, there was little popular demand for such a road and had it not been for the foresight of a handful of North Norwegians, the Highway might not be in existence today.

The history of the Arctic Highway falls naturally into four

periods. The development of fragments of roads, destined to become sections of the Highway, takes us up to the 1930s. Then follows the period of construction to join the sections into a continuous road. The German occupation saw the final completion and further modifications which were followed by the post-war era of improvement.

The Early Roads

With the exception of Finnmark, most of the early towns and villages of North Norway were on islands and almost exclusively they were maritime. Even in Finnmark, Vadsø, Vardø, Hammerfest and Honningsvåg had island sites. Their needs for roads were limited to tracks between houses and quaysides. Transport by sledge in winter and by cart in summer was undemanding. The small distances to be travelled meant that even these *roads* were little better than well-worn paths which required no more upkeep than the occasional repair of pot holes or the clearing of a ditch.

Movement by sea came naturally to a people who, by and large, gained their livelihood from fishing. Contact between neighbouring settlements was by boat. Trading centres such as Alta and Tromsø attracted their customers not by land routes but by offering suitable harbours. Provisions could be loaded into boats and carried distances which would have been impossible by road transport anywhere prior to the coming of the internal combustion engine. By using the sea, apparently isolated villages of North Norway were, in reality, more accessible one to another than similarly separated towns and villages in other parts of Europe.

The seafarers of the three northern fylker, whether on the scattered islands of the Lofotens or on the *strand-flat* of the mainland, were more mobile than their contemporaries of, say, lowland Britain up to the coming of the railway building era in the nineteenth century. Even then, the mobility of rural communities was greater in arctic Norway than in much of the United Kingdom.

The topographic and economic problems which land communications have to face in Norway have left that country lagging behind most of the rest of Europe in the matter of roads. It is true that an ancient law (circa 1100 AD) contained a provision for the maintenance of existing roads. By law, bridges had to be repaired and annual clearance of roads was demanded; trees alongside highways were to be cut back to permit the passage of a man riding with a spear lying across the pommel of his saddle! But such laws as related to ancient highways merely referred to maintenance and there was no legislation for the extension of roadways. Tracks, even in South Norway, were so poor that the country entered the nineteenth century almost roadless.

North Norway's diminutive population in 1801 was without any but the most local roads. Even large population centres such as Mo i Rana which, with Hemnes, had a population of over 4,500 in 1801, and Alta, with nearly 2,000 in the district, were served by primitive tracks. Indeed, the nearest to a system of roads or to trans-fjell communications were the *winter roads* of snow and ice such as those on the Finnmark vidda linking Kautokeino and Karasjok to each other and to the coast, or the snow roads across the Scandinavian Highlands into Skibotn and Mo i Rana. Anomalously, winter in ancient times was the period when movement was easiest and routes no longer limited to the coast or rivers. The network of winter roads, especially in Finnmark, cut across the vidder and crossed rivers and lakes. Some are still to be seen in the stone cairns which marked their paths.

Land communications thus were seasonal. In winter there was long distance travel by sledge and ski where inland settlements made this necessary. In summer, and to a limited extent in winter also, there were local parish tracks within rather than between the settlements.

Road building in Norway may be said to have started in the mid-nineteenth century with the drafting of the 1851 Highways Act. This set up Highway authorities which could draw upon State and local funds for construction and maintenance. Yet even by 1900 it was reckoned that only 6,000 miles of main road and 10,000 miles of minor road existed in a country of almost 125,000 square miles. But at least a start had been made and, with an annual expenditure of between one and two million crowns, it was only a matter of time—a matter of time in a region where, even in the second half of the twentieth century, time never seems to matter. Certainly, the North Norwegian and even more so his Lapp neighbour, lives today at a pace which can only be envied by his West European or North American cousins.

It is scarcely surprising then that North Norway was the poor relation in the road construction programmes which followed the 1851 Act. At the turn of the century the road densities were still pathetically low. Finnmark had only 0.65 miles per square mile. Troms and Nordland were little better served with 3.2 miles and 5 miles respectively. Most of these networks were little more than cart tracks. By 1915 the whole of Nordland had approximately 1,100 miles of highway designated as main road, of which 150 miles (all south of Mo i Rana) had been so classified before 1851. There were officially no main roads in the other two fylker before 1851 and in 1915 Troms had only 540 miles and Finnmark a mere 250 miles. Even these figures can easily mislead, for the term *main road* must be interpreted carefully. Roads classified in this way were under the jurisdiction of the highway authorities and met their simple standards, but, in fact, more often than not they consisted of well-maintained gravel tracks which were open to traffic probably for no more than four to six months in the year.

Part of these short lengths of main road were destined to become sections of the Arctic Highway. None but the most visionary at this time foresaw a continuous road through arctic Norway but, with difficulty, it is possible to build up a picture of the fragments of the incipient Arctic Highway as they were at the end of the first decade of the twentieth century.

At this time no more than a dozen sections were already constructed. Without exception they served well populated areas: farming communities in valleys or fishing settlements along fjord edges. No roads crossed the fjell which separated the major inlets and the longer sections of highway followed the lines of rivers and fjords.

South of Mo i Rana there was no link to nearby Mosjøen or

Sandnessjøen and travel to Hemnesøya was by boat along the sheltered Ranafjord. Northward, however, a long section of road stretched nearly 40 miles to Krokstrand, along the Rana valley. This valley provided not only the natural route but also the traffic between communities farming the alluvial and glacial soils or engaged in mining the sedimentary iron ores of Dunderlandsdalen. There was even a mineral railway built by the British-owned Dunderland Iron Ore Co Ltd to link Storforshei to the coast at Mo.

There were no insurmountable problems in the construction of this road. It had, in fact, evolved from the numerous inter-village tracks along the valley. East of Mo, the beautiful Rana and Illhullet gorges presented minor difficulties and the Langvassåga, or Blakkå, outlet of Langvatnet, had to be bridged at Reinforshei. But the attractions of Mo i Rana were an incentive to overcome these handicaps, and the town, steadily growing in importance since the mid-nineteenth century, was and still is the natural market and service centre for the valley. Beyond Reinforshei, river terraces provided a route way as well as sites for farms, and an attractive stone arch bridge allowed the road to cross the Rana again at Messingslett to follow the river along its left bank while remaining in Dunderlandsdalen.

At Storvollen, the Ranelv splits at the confluence of its two main tributaries. The western tributary leads back into the mountainous and unsettled Stormdalen and Tespdalen, while the eastern stream, Randalselva, flows through a valley indistinguishable from Dunderlandsdalen. This eastern valley, settled and farmed, carried the road as far as Krokstrand. From Mo the road had risen 1,100ft to reach Krokstrand, but even the ascents in the Rana and Illhulet gorges did not give exceptional gradients. Most importantly, the presence of a road throughout the natural hinterland of Mo i Rana made this region one of the most prosperous in all North Norway.

North of Krokstrand the forbidding Saltfjell carried no road. Not only does the fjell rise another 1,100ft above the height of the village but its southern slopes are steep, broken only by small streams. The old silver mines on the slopes of Mount Nasa had long since ceased to be worked and the occasional farm between Krokstrand and Bolna did not call for more than simple tracks and winter snow-roads. The polar circle was to remain uncrossed by a road for almost another thirty years.

The Saltfjell drops more gently on its northern side. The watershed between the southerly and northerly flowing streams is nearly 2 miles north of the Arctic Circle at Stødi and from here the winding Lønsdalen carves a natural path down through the plateau. All the same, even though less steep than the southern side, Lønsdalen drops almost 1,900ft in 15 miles and, in 1910, this upper stretch of the valley was without a road. But, at Hestbrinken, a road led east into Junkerdalen and the main road continued northwards a full 30 miles through Saltdalen to the head of Saltdalsfjorden at Rognan. This was an important section of road whose path is almost unaltered in today's Arctic Highway. A very gentle gradient carried the road through farming country and, especially north of Saltdal, the valley had been settled for centuries and become renowned for its boat building. Saltdalen must have had well trodden paths along much of its length for hundreds of years.

The route around Saltdalsfjorden towards Fauske was too difficult for the construction of a road, the mountains dropping almost vertically into the fjord. Instead, the journey could be undertaken by boat and Finneid or Fauske reached in half a day. The whole Saltfjord was well served by local craft, yet it had no road connections between what were already important settlements. Bodø at the mouth, approached by way of the treacherous Saltstraumen, the rich Misværdalen, the mining-market town of Fauske and the port of Finneid—all were unconnected except by water. Finneid was in an especially strange position for it was served by a railway but had no main road. This little port had become the outlet for the Sulitjelma copper pyrites being exploited in the mountains by a Swedish company. A 0.75m gauge line, originally intended only for ore, was by now carrying passengers from Sandnes to Sjønstå. Boats were then used to complete the journey across the lakes, Øvrevatnet and Nedrevatnet, east of Finneid. It is interesting to note that although, originally, a road was built to carry the copper ore from the mines to the lakes, it was soon replaced by the railway in 1889 and it has since been allowed to decay.

Only two miles round the fjord from Finneid is Fauske and a road linked this small town with the head of Sørfolda, about five miles further north. In fact, more than one road had been built in the broad and comparatively flat Fauskeidet depression, for this valley contained valuable marble quarries and timber as well as farmland. Its marshy surface had been a problem, especially in spring, but the watershed is at only 250ft and the road here had become an important section of highway between two fjords.

Beyond Sørfolda there were no roads for the next 125 miles of the route north. Nothing but old farm and village tracks skirted the deeply indented coast or reached back into the valleys; nothing at all crossed the plateaux between the fjords. Sparsely inhabited, this polar part of Nordland, comprising the *distrikt* of Nord-Salten and much of the *distrikt* of Ofoten and Lødingen, was almost as starved of roads as was interior Finnmark. Had it not been for the presence of Narvik, the Arctic Highway's largest town, no main road would have been built south of the Nordland-Troms border.

However, in 1910, Narvik was beginning to reap the benefits of the construction of the Ofoten railway completed eight years earlier. The population already exceeded four and a half thousand whereas twenty years before only a handful of farms had occupied the site. The settlement had spread beyond the peninsula separating the Beisfjord in the south from the Rombaksfjord in the north to Grindjord and the head of Herjangsfjorden. A road was built to link Grindjord to Ankenes around the edge of the Ofotfjord and a ferry connected Ankenes to Narvik. Narvik's roads were well maintained and it was possible to cross Rombaksfjord by ferry to Øyjord. From here a road ran round the shore of Herjangsfjorden to Bjerkvik.

Important though these roads were to the greater Narvik

community, they did little in their short six or seven miles to contribute to regional communication. North of Bjerkvik, just as south of Grindjord, no road had been built to join Narvik to its natural hinterland. Narvik was to grow up as a Swedish outpost rather than as the natural capital of Nordland which a road system might have made the town.

Northward, behind Bjerkvik, the mountains rise steeply to the Troms-Nordland border. No easy routeway links Herjangsfjorden with Gratangen and for many years the Gratangen Ranges were to prove too difficult for a road. North of Gratangen, the Sølvfjell had also proved impassable and even Sangsdalen lacked a main road. But, from Brandvoll, where a marshy depression separates Sangsdalen from Bardudalen, the picture changed dramatically. By 1910, this latter valley and that of Målselv, of which it is a tributary, accommodated roads which were at least the equal of any in North Norway.

These valleys had been settled since the end of the eighteenth century when flooding in Østerdalen had led to northern migrations of farmers. Here they found a climate and topography which allowed cultivation and pastoral activities on a scale unparalleled elsewhere in North Norway. Despite a certain loss of population in the mid-nineteenth century through emigration to North America, these valleys contained the largest and most widespread farming community north of the Arctic Circle. The villagers had built roads alongside the river courses in the broad valleys. Tributary streams were crossed by wooden bridges and the wider Målselv had a ferry at Fredriksberg. Those sections of road which were eventually to become part of the Arctic Highway kept fairly close to the river banks for most of their lengths, but often these were alternative parallel roads on higher and drier ground which could be used during spring floods. The Bardu and Målselv had no less than thirty-five miles of road which was to be incorporated into the Arctic Highway. Moreover, the importance of the valleys had a singular bearing on the final path of the Highway, causing this route to be chosen rather than a coastal road through Lavangen, Salangen and Finnsnes.

The thirty-five mile stretch of highway was separated from the next northward road only by a narrow roadless col of ill-drained, lake-strewn plateau rising to over 700ft at its lowest point. This is Heia hill and, although by no means an exceptional obstacle in the path of a road, it was effective enough to give a ten mile break in communications.

A forested path led down through Sagelvdalen from Heia but a main road was not to be found until Storsteinnes. From this small but ancient trading village on the Balsfjord, a road had been constructed through the intensely farmed country round the head of the fjord, across the Nordkjoselv and eastward over a low pass to Storfjorden which is the southern arm of the great Lyngenfjord: a total of over twenty miles of highway. The construction of the peripheral Balsfjorden road is accounted for by the relatively large number of farms around the fjord between Storsteinnes and Nordkjosbotn, but the inter-fjord link is differently explained.

Between Nordkjosbotn and Storfjord a natural route is cut by the Nordkjoselv in the west and the Oterelv in the east. This narrow neck, Balsfjordeidet, overshadowed by mountain peaks rising over 4,000ft to the north and to the south, scarcely reaches 300ft at the watershed and forms one of the most beautiful natural passes in all Troms. To build a road through this pass was an opportunity not to be missed, linking, as it did, two important fjords, namely Bals and Lyngen.

Lyngenfjorden had been settled for many centuries before the present. There is evidence of a stone-age site on the western shore and for hundreds of years the fjord had been a centre for trade and cultural exchange between Norwegians, the Kvaens—or Finns—and the Lapps. The area was a Norwegian outpost by the ninth century, and by the middle ages the fjord's shores had become part of the routeway between Tromsø and what is now Swedish Lapland. Tracks of ancient origin lead from the mainland opposite Tromsøya, through Breivikeidet, to the shores of Ullsfjorden. Boats were used to cross the fjord and reach Kjosen at the western end of a narrow isthmus behind Lyngseidet on the

Lyngenfjord. The journey could then be continued either by boat across Lyngenfjord or by track around its head to Skibotn and thence over the mountain pass and into Finland.

Despite the existence of these old tracks, only a short section of main road had been built along Lyngenfjorden by 1910. The highway stopped short at the fjord head, after passing through Balsfjordeidet, and the only other road was from Arøybukt to Kvalvik, about three miles either side of Lyngseidet. One deterrent to road building was the risk of avalanche, especially from the precipitous slopes of Pollfjellet and Rastebyfjellet which tower over the fjord edge, but the dependence on water transport, as so often elsewhere, must have lessened the need for a continuous road.

Today, the Arctic Highway is broken by a ferry crossing between Lyngseidet and Olderdalen on the eastern shore of the Kåfjord arm of Lyngenfjorden. Sixty years ago there were almost no main roads whatsoever north of the Lyngenfjord. The only section of the Arctic Highway in existence at this time was the Tanafjorden-Varangerfjorden link. The Tana river had been used for summer traffic into interior Lapland ever since man had ventured this far north and its valley and the northern side of Varangerfjorden have been settled for longer than any other part of North Norway. The primitive tracks had been improved and by 1910 there was a continuous section of highway from the head of the Tanafjord through to Karlebotn at the western end of Varangerfjorden. In fact a road ran almost the whole length of the Varangerfjord, through Vadsø and on to Norway's most easterly point, the coastland opposite Vardøya. The road from Varangerbotn to Karlebotn was really a branch of this highway.

The thirty-mile Tana-Varanger road was open for only a few months each summer but the same route was used in winter by sledge and ski when the frozen Tana river could easily be crossed and the ferry at Seida became unnecessary. The banks of the Tana provide an easy route along the broad floor until the road leaves the valley at Skipagurra. Here, the Seidafjell divide, although rising to over 400ft and with a steep easterly descent, had been crossed by countless generations of migrating Reindeer Lapps and it was this path that evolved into the Highway.

Elsewhere in Finnmark there were only four fragments of road which were eventually to be part of the Arctic Highway. Kirkenes had just become important as the outlet of the rich iron ores from Sydvaranger and a short main road had been built westward from the town for a couple of miles. At Alta, the still distinct settlements of Bossekop and Elvebakken were linked to Rafsbotn round the fjord-head, a distance of ten miles. This mostly farming and trading community was quite isolated from the rest of Finnmark and its affinities were with the vidda and Finland to which it was connected by snow roads.

The head of Porsangen was similarly linked with Karasjok in the interior but, because of its fishing activities, the fjord, on its western side, carried a road from Lakselv to Kolvik. This twenty-mile section, although classified as a main road, was exceptionally poor and the numerous bridges over streams flowing into the fjord were often in a bad state of repair. Moreover, the mountains rise steeply from the shore in the north and avalanches, rock falls and melt water annually caused havoc on the track which passed for a road.

In west Finnmark, across the boarder with Troms, there was built in 1910 the last main road which was to become a stretch of Norway's Arctic Highway. This little road, only five miles long, crossed the neck of the Øksfjordkollen peninsula by way of the low-lying Alteidelv valley. It joined the old trading post of Alteidet with Nordshov and Langfjordbotn. For over a thousand years the route had been used by reindeer herders, both Norwegian and Lapp, in their spring and autumn migrations.

Such was the Arctic Highway, or rather the Arctic Highway to be, at the end of the first decade of the twentieth century. Patches of gravel roadway, none more than forty miles in length, and a total of perhaps two hundred miles. Two hundred miles of a highway which needed over nine hundred miles if Mo i Rana and Kirkenes were to be linked. It must also be borne in mind

Page 51 (*above*) The Nordland railway crosses the Highway near Storvollen. The road surface here is good oil-bound gravel. The railway was opened as late as the early 1960s. (*below*) The new Highway and old bridges at Sondli. Sections of the old Highway like that near the old bridge often provide sites for campers in summer.

Page 52 (*above*) The Polar Circle café on the Saltfjell. The plateau here is open and bleak, even in summer. The Highway across Saltfjellet was kept open in winter for the first time in 1968-9. (*below*) The Highway and wooden railway tunnel north of the Polar Circle. The tunnel prevents snow from blocking the line. The Highway's solution to the same problem is a raised surface.

that these two hundred miles were built along the easiest parts of
the route, that no major fjell had yet been crossed or fjord
bridged, and that the standard of construction satisfied eighteenth
rather than twentieth-century requirements.

Amazingly, the picture hardly altered in the next two decades.
By 1920 the only additional section of main road to have been
completed was a four-mile length from Lakselv to Stråskogen at
the head of Porsangen. In 1930 two more extensions had been
built. The first completed the Highway along the western shore
of Porsangen crossing the necks of the Veidneset, Kovholmen and
Trollholmen peninsulas to join Kolvik to Russenes. The second
was just three miles long on the Kåfjord to meet the needs of
the local community.

Of course there were other roads lying off the Arctic Highway's
path and there were countless miles of track through forests,
across wild open plateaux or along fjord shores, some of which
followed routes the Arctic Highway was to take. Nevertheless,
the problem of building a continuous road, even with short
ferries, from the Finnish border in east Finnmark to Mo i Rana
just south of the Arctic Circle was a daunting one. But it was a
problem that had been accepted and the concept of an Arctic
Highway was no longer the visionary's dream but a policy of the
State.

From the early 1920s, surveys for the new road were in hand
and route planning was under way. The general line of the link
sections seemed fairly clear on a small scale map, but the details
of routeing, the choice of a valley side, the selection of a ferry
point or decisions concerning bridging sites, were matters to be
resolved only after considerable ground survey and painstaking
analysis of the data collected. To have transformed the ancient sea
highways into land roads would have been to build a coast route
but, although this would have served the greatest number of
communities, it would have been impossibly long. Instead, the
challenge of trans-fjell sections had to be accepted and the river
valley and coast sections of the Highway already in existence
had to be linked by roads across the high plateaux. In accepting

D

this as the principle on which the Arctic Highway's route was to be planned it also had to be acknowledged that the road would include considerable stretches which would be closed in the long winter.

The bulk of the survey was accomplished by 1930 and the next decade—or eleven years, to be precise—saw the completion of the Highway.

The Highway Completed

In 1932 it was estimated that rather more than 600km, or about 400 miles, of the projected Highway's route was completely without even the most primitive roadways. Work on the Highway was hampered by lack of funds and, even more, by the very short summer season in which construction could take place. The repair of frost and vehicle damage to sections already built frequently took valuable time and capital away from new works. Yet slowly the gaps were being closed. By the mid-30s the only unplanned route lay between Karlebotn and Kirkenes and the greatest obstacles preventing the completion of the road were the high fjell, notably: Saltfjellet, Sjettevassfjellet, Kvænangsfjellet, Sennalandet and Børselvfjellet.

Perhaps the most important sections of Highway that had been built were in South Troms. In 1935 it had become possible to travel by road from Grindjord to the Straumenfjord well north of Lyngenfjorden. The mountainous Gratangseidet beyond Bjerkvik had been conquered by a winding road across the Troms-Nordland border and the Narvik and Bardudalen communities could communicate along the Highway—except, of course, in winter. The break in the Highway between Målselvdalen and the head of Balsfjord no longer existed now that a road replaced the forest path and, at last, the Lyngenfjord had a continuous road along its western shore. The crossing of this fjord was, at the time, between Årøybukt and Djupvik but a road was already projected between Djupvik and Nordmannvik and the latter was eventually to become the ferry point. Northward the

Highway continued to about half way along the eastern shore of Straumenfjorden passing through the many fishing settlements that cling to the *strand-flat*.

Other stretches completed by the mid-30s included a short length from the head of the Sørfolda fjord to the village of Sørfold on the western side of the inlet and a fifteen-mile section between Ulsvåg and Korsnes. In Finnmark little work had been done. The Highway north of Alta was extended only a few miles beyond Rafsbotn, although the links with the interior had been strengthened by a road from Gargia to Masi almost reaching Biggeluobbal. The only significant part of the Arctic Highway to be constructed in Finnmark in the early 1930s was between Skaidi and Olderfjord, a matter of fourteen miles.

It was left to the second half of the decade for the really difficult gaps in the Highway to be finally closed. The preliminary work had taken nearly fifteen years and the resultant road was, ironically, to be almost finished just in time for the occupation of the country by the German army.

The opening of the Highway across the Saltfjell in 1937 by the King of Norway was a proud moment in North Norway's history. The Arctic Highway now linked two of the most important towns in Nordland: Mo i Rana and Bodø, and the Polar Circle had finally been crossed. Although ranking as a major landmark in the Highway's construction, the Saltfjell road clearly illustrates some problems as yet unsolved. The plateau rises well over 2,000ft and winter conditions are made the more severe by its interior position. The lowest part of the plateau had to be avoided because it acts as a snow trap and the sides of the dissecting valleys offered the best route. After much deliberation the path of the Highway was chosen, only to prove to be on the most thickly covered snow slope throughout winter. The road could be kept open only from late May to October. Had the engineers heeded the advice of the Lapps who seasonally use the plateau for grazing their reindeer, the other side of the valley would have been chosen and the snow problem reduced. Too late for the road the difficulties were realised and it was left to the railway to follow the better route. The road remained closed in winter up to 1967-8 but cars could be transported by rail between Mo and Saltdal.

Two years after the official opening of the trans-Circle section almost the whole of the Highway had either been built or was in an advanced state of preparation. The high Kvænangsfjellet took the road up to 1,300ft and through the Lapp summer pasture land. The steep approach on the southern side had proved difficult and, in 1939, the northern fjord-side road to the ferry point at Sørstraumen had not quite been completed. Further east, the road between Karlebotn and Kirkenes, one of the longest sections, of seventy miles length, had been opened to traffic.

This east Finnmark road had proved to be a Herculean task in creation and maintenance. Much of the route lay away from the coast on high ground or else it was exposed at fjord heads open to northerly winds across Varangerfjorden. The traffic density, even in summer, was especially low and this road's construction might be considered as symbolic, the last link in the communication chain, rather than as serving any very useful purpose. The Highway here was built to a very low standard, with parts scarcely wide enough to take a single vehicle. The local fishing settlements kept to their traditional sea roads and Kirkenes was more interested in shipping its iron ore than in an Arctic Highway which hardly merited so grand a title.

By the time much of the rest of Europe was becoming embroiled in World War II, there was only a little more work to be done on the Highway. The major missing links in the chain were the wild, uninhabited Sennalandet and Børselvfjellet in Finnmark, and Mørsvik and Kråkmo in Nordland. These three plateaux are at once high and little settled. The Senna plateau rises over 1,200ft, Børselvfjellet reaches to over 600ft, and passes between Mørsvik and Kråkmo not only exceed 1,200ft but are steep and without a clear route. All the same, when German forces entered the country in the spring of 1940 they found Norway's Arctic Highway complete except for the finishing touches which they were soon to put to a road which they considered of vital importance.

The Highway in War

The history of the invasion of Norway in April 1940 has no place in this book. Nevertheless, from almost the first days of invasion and occupation, the Arctic Highway played a prominent rôle in the war and, later, German strategy demanded that the road not only be maintained but improved.

In those early days of what has become known as the Narvik Campaign, British and other Allied forces had reason to be thankful for the Highway's existence. The detachments of Scots Guards sent to Norway in May 1940 to retake and hold Narvik—considered necessary by the Allies because it was the outlet of Swedish iron ore—found themselves in full retreat northwards from Mo i Rana. They had arrived too late to halt the German advance south of Mo, even though the lack of a road to Mosjøen was to their advantage. Outnumbered, ill-equipped and, especially, ill-prepared for an unusually cold spring, they could do little but delay the German army as it pressed towards Narvik along the Highway. Their commander, Col. Trappes-Lomax, later to be *sacked* for his failure to do the impossible and halt the enemy at Krokstrand, led his dwindling force north across the new Saltfjell road over the Polar Circle. In a week the Guards covered the 100 miles of Highway, blowing-up bridges as they went, and reached the village of Rognan. This was then the ferry point to Langset from whence the Highway continued north through Fauske. Eventually, at the end of May, what remained of the battalion was evacuated from Fauske and Bodø. Without the road it is doubtful if any of these men would have escaped capture or death.

The abortive campaign had left Narvik in ruins following a continuous air attack on 2 June, and by the end of the first week in June the whole Norwegian battle was over. There followed five years of an occupation which, although little has been written of it, was one of the most cruel and inhuman of the whole war.

Initially the Germans' task was to rebuild the Highway's bridges which had been destroyed either in the Allies' retreat or during the fighting. Most of the old bridges were fine masonry or steel girder constructions. These, now shattered, were usually replaced by strong wooden bridges often using the original abutments which still stood. Such was the case of the Messingslett bridge where the 120ft span had to be renewed. Similar large wooden bridges were built at Reinforsheien and at Krokstrand. In other instances, such as the crossing over the river Junkerdal, more elaborate suspension bridges, at least the equal of the old spans, were built. Further down stream on this same river the Germans put up a three-span girder bridge, an exact copy of the Norwegian design.

The broken bridges were largely between Mo i Rana and Fauske or around Narvik. Elsewhere the Germans had other plans for the Arctic Highway. North Norway was now completely under occupation and Germany, never slow to appreciate the importance of communications and supplies to her forces, looked upon the Highway as the vital link between her scattered troops. Moreover, as the war progressed, the German High Command began to think in terms of an Allied attack through North Norway. These fears were probably first aroused by Commando raids on the mainland and islands of the north but, as Russia entered the war, Germany became more and more convinced that this arctic front was of primary strategic importance. Thousands of Germans were billeted on reluctant Norwegian hosts and the job of making the Arctic Highway serve the German army got under way.

First the partly constructed sections had to be completed. The plateau gaps had to be closed in order to allow continuous traffic if the Highway was to act as supply line. As has been said, most of the preliminary work had been done and by mid-summer 1941 the Arctic Highway was finally completed. With ten relatively short ferries it was possible to take a motor vehicle over 900 miles from Mo i Rana to Kirkenes. With misplaced pride the occupy-

ing forces held an opening ceremony and erected a plaque where
the Highway crosses the Polar Circle on the Saltfjell. For the
local people, recalling their own King's opening of this section in
1937, it was a sad occasion. How perverse of fate to decree that
North Norway's great road should be completed in such unhappy
circumstances. It was, and is, a Norwegian achievement of which
the people are justly proud. They are equally justly angered by
tourists from Germany who today, returning to the arctic with
their families, point to the Highway as the road they built. Nothing could be further from the truth, for not only was the route
planned before the occupation but all the primary engineering
had been done. Moreover, the Germans soon came to rely not
on their own military engineers but upon prisoners of war to work
on the Highway.

Once the road was complete, the Germans set about improvements to raise the standard of the poorer sections such as those
along Varangerfjorden and south of Narvik towards Røsvik.
Further, they aimed to reduce the number and length of the
ferries and to keep the road open during the winter. For all this
work they brought many thousands of prisoners into the region.
In Finnmark and North Troms Russian labour was used, but
in Nordland and South Troms the workers came from Central
Europe, largely Yugoslavia and Czechoslovakia. This forced
labour will for ever be remembered by those Norwegians who
witnessed it.

In June 1942 three German troopships arrived from Stettin
with some 8,000 Yugoslavs. These prisoners had been rounded-
up with little warning and with no indication of their destination.
Not unnaturally they had put on their best clothes and carried
what they could of their personal possessions. They were not to
know that they were to spend the next months, for as long as they
could survive, north of the Arctic Circle.

Arriving in summer, conditions were not at first unbearable.
Many were taken to Saltdalen to improve the trans-circle road
and to complete the Highway between Rognan and Langset and
so make the ferry across Saltdalsfjorden unnecessary. Housed in

crowded camps, almost without food and with totally inadequate clothing, their position gradually worsened as the short summer gave way to the long arctic winter. Disease, especially TB, claimed hundreds of lives, others died of malnutrition, were executed or died from constant beatings. The local people did what they could to help and there are many stories told of bravery by Norwegians, especially the women, who risked their own lives to take food to the prisoners. Less than 1,000 of the Yugoslavs survived and their dead comrades today share a cemetery at Botn with some of the Germans who were their captors. The Arctic Highway became known as the *Blood Road*.

The efforts of the Germans to keep the Highway open over the Saltfjell during winter never succeeded and the Slavs died in vain. Equally unsuccessful were the German attempts to maintain other parts of the road clear of snow. Sennalandet, the Kvænangsfjell, Børselvfjellet and the most easterly part of the Highway could not be kept open despite renewed efforts and more vigorous attempts each successive winter. It had never been the Norwegians' intention, in planning the route of the Highway, that it should be an all-weather road.

Greater success was had in reducing the ferry crossings. Mention has already been given to the elimination of the Rognan-Langset ferry. Work had already started on the road round the head of the Saltdalsfjord before the German occupation. It was only a short ferry of some three miles, but it took twenty minutes and was a hindrance to through traffic. The road which replaced it was hewn out of the mountains at the fjord head and was a useful contribution to the Highway's service to this particularly heavily settled region.

Further north, Bognes was to replace Korsnes as the ferry point for Skarberget; on the Lyngenfjord the old ferry between Arøy-bukt and Nordmannvik was supplemented by a larger one further south, between Lyngseidet and Olderdalen, as soon as the road on the eastern side of the fjord was complete.

North of Kvænangsfjell, the ferry between Sørstraumen and Badderen across Kvænangsfjorden became unnecessary when the gaps in the Highway through Kvænangsbotn were closed. In summer the Tana was spanned by a wooden pontoon bridge near the Seida ferry, but in winter the bridge was removed and traffic crossed the frozen river.

Throughout the occupation, the Highway was guarded by patrols and by strategically positioned gun emplacements. Many of the guns were mounted at points where artillery could both defend the road and be used against attack from the sea. Others were sited such that they could command the Highway where it passed especially narrow defiles. Telegraph stations along the road were turned into command posts and the guards spent many a long month in an alien environment in which the winter darkness was as much an enemy as the Norwegian partisans. In fact, the Highway never came under attack except by the elements.

Although the Germans succeeded in keeping the Highway open for longer periods in winter than had been necessary in peacetime, they were defeated in their attempt to use it as a through road for twelve months in the year. Their own forces, thousands of prisoners from the labour camps, and all the advanced technical resources of the German High Command, were powerless against the winter snows and spring melt water. Even the erection of wooden tunnels and avalanche sheds over the road at the most vulnerable points was of little use.

The collapse of the German forces in Norway in the autumn of 1944 saw a ruthless scorched earth policy applied by General Rendulic. German troops stationed in Finland were in general retreat northwards following the armistice of 19 September. With the Russians hard on their heels, they crossed into Norway from the Petsamo corridor and continued westward through Finnmark taking the German army in Norway with them. The speed of retreat was too rapid at first to allow much destruction, but in west Finnmark and Troms everything that could be burned or blown up was destroyed.

The Arctic Highway played its part in the retreat by acting as the route by which heavy equipment could be transported. It was also used to bring the local people to the ports to be forcibly evacuated as their homes were burned and their livelihoods ruined. Fearing that the Russians would also use the Highway and so close on their retreating forces, the Germans demolished bridges, dynamited embankments and threw barriers across the road in the wake of their own withdrawal. No less than 350 bridges were destroyed in Finnmark and North Troms and the Arctic Highway was left without a single major span which had not been completely wrecked or seriously damaged.

The policy of retreat and destroy was applied as far south as the Lyngenfjord, although the Russian advance did not extend west of Seida. By May 1945 the war was over and North Norway and its Arctic Highway were to rise, phoenix-like, again.

The Post-war Period

With the return to more normal conditions when peace came to Europe the most immediate need of the Arctic Highway was the repair or replacement of bridges and of ferry quaysides destroyed in the German retreat. It was estimated that in Finnmark alone the cost of repair of buildings and roads was 50m Nkr. There was no intention, however, simply to allow the reconstruction programme to stop there, and the quarter century since the war has seen improvements to the Highway in three principal ways:

i. In the reduction of ferry links both in distance and in total number.

ii. In the standard of the road in terms of widths, surfaces, curvature of bends, gradients, etc.

iii. In the battle to keep as much as possible of the Highway clear for traffic even in the middle of winter.

Many of the pre-war and wartime bridges were below the standard needed for the increasing traffic on the Highway after the war. Temporary bridges, hastily erected as the evacuated population returned to the North, soon needed replacement and even those bridges which had survived the retreat-and-destroy strategy were often out of date. The number of bridges on the Highway is, of course, enormous. In addition to the crossings at or near the heads of fjords, the fact that the Highway often runs along the coast means that innumerable outlets need to be bridged. Even on the vidder the density of streams incised into the plateaux necessitates frequent short span links. A policy of bridge improvement has been costly but successful in the last twenty-five years.

Many of the new bridges are, naturally enough, in Finnmark. It was here that almost total destruction had taken place and between 1947 and the early 1960s the opportunity was taken to improve links in the building of well over a dozen new major bridges. The alignment of some of the older bridges was frequently dictated by the need to reduce the span to a minimum. This in turn led to difficulties on approach sections of the road which was taken out of its way in order to reach the bridging point. A new bridge, wider and with a longer span, allowed a more rational route to be taken by the Highway, even if the displacement was only a matter of a score of metres.

The 1952 Salletjokka Bru on the western side of Porsangen is just such a case where the old bridge, now collapsed, and its approach roads are bypassed by the realigned Highway. Likewise, outside the *suburb* of Kirkenes, Hesseng, the new river crossing is only a few metres down stream of the old, but the improved approach is clear for all to see in the remains of the old road.

In the late 1940s four new bridges were constructed over rivers flowing into the Porsangerfjord alone: Børselv Bru and Karijokka Bru in 1947; Ytre Billefjord Bru in 1948, and a year later the Stabburselv got its new bridge to replace the pre-war 164-foot single span lattice girder bridge. In the 1950s, along this same fjord, another half dozen or so new bridges were built, the last being Brennelv Bru in 1958.

Most of these bridges are of a simple reinforced concrete platform construction, about six metres wide and with only a low stone wall about a foot high at the sides. Occasionally a tubular steel handrail is added and, on longer spans, a lattice girder type bridge is used. Each bridge is the essence of simplicity. Only one really large bridge needed to be built in Finnmark. This was the link across Norway's third longest river, the Tana. Tanaelva is not only long but wide and its current fast. The former ferry and pontoon bridge between Vestre Seida and Seida was one to two miles down stream of the point where the channel narrows and the 1948 bridge was built. The modern suspension bridge is the second to be built at this site and the new bridge, 180m long, allows the Highway to proceed uninterrupted by ferries throughout Finnmark.

While in Finnmark there were still some ninety bridges needing replacement and a quarter of the German damage remained unrepaired as late as 1955, the scorched earth policy of General Rendulic left most of the bridges of the southern two fylker, Troms and Norland, intact. Even today, basically pre-war bridges such as Målselv Bru at Olsborg and Vollan Bru (both built in 1939) serve the Highway adequately. Less adequate are the vidde bridges over tiny streams, where pre-war widths of less than four metres cannot cope with the heavier summer traffic on sections such as that crossing the Polar Circle on the Saltfjell. However, bridges built in the southern fylker have usually been part of schemes aimed at reducing the ferry sections of the Highway, or replacements of the wooden bridges built by the German occupying forces.

Despite its short life, the Highway has often seen as many as four bridges erected successively at the same point. The destruction of the pre-war bridges in first the Allied and then the German retreat, hasty improvisation of temporary bridges and their ultimate replacement: such is the record. Typically, the bridge over Saltdalselva near Pothus, just south of Røkland in Nordland, had its third completely new bridge after the war. The present two-span steel girder structure replaces the three-span lattice bridge which the Germans erected to the same design as its Norwegian predecessor.

The most spectacular bridge completed since the war has been the Rombak Bru north of Narvik. Until the building of this giant suspension bridge, it was necessary to cross the Rombak arm of Ofotfjorden by ferry. Although the crossing (between Vassvik pier and Øyjord) was short at the narrow entrance to the fjord, and the ferry took only twenty to twenty-five minutes, it was an especially irksome break in the Highway. Narvik is the largest and most important town on the road north and its isolation from its natural hinterland remains an unsolved problem. Situated on a peninsula near the head of the fretted coastline of Ofotfjord, the town is the product of the Ofot railway, built in the 1880s to link the ice-free Norwegian coast with the Swedish iron ore fields at Kirunavarra. Today it is still something of a Swedish outpost, but the Arctic Highway is its link with the rest of Norway. Furthermore, the districts to the immediate north and south of Narvik are well settled and the Highway, northwards, also provides access to Route 19 and the Lofotens. The Rombak bridge is, therefore, important not only to the Highway as such but of special significance to Narvik.

To build a bridge at the mouth of the fjord, at the ferry point, would have been impossibly expensive. It was necessary to look for a narrower part of the inlet and this was found about nine miles east of the town where the fjord contracts into Rombaksbotn by way of the Rombakstraumen. Here, between 1961-4, was built the Rombaksbru, the largest suspension bridge in Norway. The new approach section of the road from Narvik skirts the edge of Rombaksfjord well above the sea. This part of the Highway is built to a high, 1990s standard, with an asphalted surface, and its elevation ensures that the winter's snow is unlikely to cause it to be closed for any length of time. This is especially important on this north-facing slope and the height of the road also brings it level with the bridge without a steep gradient approach.

The Rombak bridge has a total span just short of 2,500ft with a central section, between the towers, of 1,066ft. The centre of the bridge is nearly 150ft above the fjord and the northern and southern towers rise about 250ft and 290ft respectively above sea level. Although subject to strong winds, especially in winter, the bridge crossing affords wonderful views westward to the mouth of the inlet, making it tempting to stop. The cost of construction is being recouped by toll charges made at the northern end of the bridge. The income is considerable at about twenty-five English new pence for each car and proportionately higher charges for larger vehicles, gaining from the phenomenal increase in traffic since 1964. The bridge's opening also brought new life to the territory north of the fjord. Bjerkvik, at the head of Herjangsfjorden and at the junction between the Highway and Route 19, has become, practically speaking, a suburb of Narvik despite the fact that it is twenty-three Highway miles out of the town.

Just south of Narvik, the Beisfjord bridge now saves the ten to fifteen minutes the old ferry used to take. This bridge allows for the passage of ships into the fjord but effectively links suburban Ankenesstrand to industrial Narvik. However, in the next hundred miles of the Highway, south toward Mo i Rana, there are still three ferries breaking the continuity of the road. One of these crossings, over the Leirfjord arm of Sørfolda, has been considerably reduced since the war; another, at Grindjord, is to be eliminated (see Chapter 8) and in 1969 what was a fourth ferry was bypassed by a bridge.

The Arctic Highway north of Fauske was originally built along the western side of Sørfolda to Røsvik. This route provided the easiest path, for, although the relief is steep and rugged, the western side of the fjord, in contrast to the east, is not broken by inlets. However, the road terminated at Røsvik and a linking ferry carried vehicles eight or nine miles up the Leirfjord whose steep sides drop precipitously into dark waters. The journey, to Bonnåsjøen, took over an hour and included the difficult passage across the open neck of the Y-shaped fjord.

At a cost of 58m Nkr a new section of the Highway was built from Vargåsen round the eastern side of Sorfolda and the southern side of Leirfjorden to Sommerset. The distance is only forty km, or twenty-four miles, but it is one of the most modern stretches of the whole Highway, both in concept and construction. The whole surface is of asphalt and built to standards which will comfortably cope with the demands of the 1980s and 90s.

This superb piece of engineering is a joy to drive on and, with its graceful bridges and gentle curves, the Highway adds its own beauty to the splendid natural scene of fjord and mountain (*see plate, page 34*).

The new route has to contend with a broken coastline and steep-sided mountains. To overcome these natural difficulties, the Highway requires two large arch bridges and no less than six major tunnels, as well as two or three minor rock arches and short tunnels (*see plate, page 34*). Furthermore, the road has to climb to over 300ft before it can finally drop to sea level to reach Sommerset which, since 1966, has become the new ferry point Bonnåsjøen. The shorter ferry journey takes only a quarter of the time of the old and, quite apart from this saving in time, the magnificent new road, which is described in the next chapter, is full of interest. Eventually even this short ferry will be bypassed by the Highway (see Chapter 8).

One hundred and twenty miles north of Bonnåsjøen is the newest piece of Highway, built to excise yet another ferry. Here, on Efjorden, the situation is very different from Leirfjorden. The old ferry crossed the inlet between Saetran and Forså in about twenty-five minutes but, because of the steepness of the fjord walls, the ferry points were not on directly opposite shores. Instead of a short crossing, for the fjord is quite narrow, the ferry boat had to travel nearly four miles up the inlet to reach Forså. Fortunately, two rocky islands lie off-shore from Sætran almost blocking the entrance to the fjord. These have been used as stepping stones by the new section of road which now crosses the fjord to a point directly opposite Sætran by way of these two islands and three bridges. The spans increase northwards and consequently three different types of bridge have been used: steel girder platform, arch and suspension.

Two particular problems delayed the opening of this new part of the Highway. Difficulties were encountered in the building of the approach road on the southern side. Retaining walls were needed to hold back lacustrine deposits of sand which would have wasted on to the road. Then it was discovered that the cables delivered for the longest span, the northern suspension bridge, were the wrong specification and the opening, planned for 1968, had to be put off for a year.

On the northern side of the Efjord is a new eight-km section of road cut out of and into the mountainside. The surface is asphalt and at the eastern end a tunnel has been cut to reach the old Arctic Highway just beyond the Forså pier. The two ferry boats, *Solfrid* and *Frydenlund*, are no longer needed.

No major re-routeing of the Highway has taken place since the war and it is doubtful if any such alteration will ever be made unless it be necessitated by new bridges across the larger inlets. Instead, there has been a programme of straightening dangerous or difficult curves and eliminating sharp bends. No one fylker has received more than its fair share of the money available, although the mountainous nature of Troms and Nordland have made this type of improvement more necessary there. Some of the work has been carried out in conjunction with other modifications to surface, weight tolerance limits, bridge maintenance and snow-trap ditches.

The improved surfaces over the last half decade are immediately evident to the traveller in North Norway. What were poor surfaces requiring especially careful driving and low speeds even in 1965 are today often the equal of any comparable motor road in Europe. The Highway may be no *autobahn* or *autostrada* but its new surfaces are gradually allowing a combination of speed with safety.

Three types of surface are currently in use: water-bound gravel, oil-bound gravel and asphalt. In recent years the water-bound gravel has been replaced by an oil-bound composition on almost all the newly surfaced sections and the lengths of the asphalt segments have been increased. The criteria used in deciding the location of improvements to be made to surfaces fall roughly into two categories, namely: the demands made by increasing traffic densities, and the necessity to improve sections of the Highway which, for one reason or another, have been neglected and are deteriorating or are particularly susceptible to the weather.

Page 69 The Highway alongside Saltdalselva—a particularly quiet and lovely valley in summer.

Page 70 (*above*) This view from the Highway near Nordreisa is typical of the northern fjords and contrasts with the narrow steeper inlets of South Norway. (*below*) The Highway crosses the bleak tundra of Ifjordfjellet. Small snow patches, shallow lakes and the absence of trees are typical of the extreme north.

If we examine those census figures that are available for the Arctic Highway, the general trends outlined in Chapter 1 are confirmed. Traffic density is increasing at a rate that might be described as alarming and there is the usual pattern of heavy concentration near the larger settlements and between towns and villages which are in reasonably close proximity. (Details of the 1965 Census are given in Appendix B, Table VI.) More significant still are percentage growth rates of traffic along the Highway. In seven years, between 1960 and 1967, traffic in the region of Andselv increased by over 170 per cent and through Vollan by 120 per cent. Even at the head of the Langfjord arm of Altafjorden increases in the same period amounted to nearly 100 per cent in a southerly direction and 56 per cent towards the north. Everywhere the story is the same; that of annual increases in motor traffic in excess of 10 per cent per annum and often reaching 15 per cent per annum. In the latter part of the decade the trend has been an ever larger percentage increase in each successive year. When these percentages are translated into numbers of vehicles, the problem of coping with the volume of motor vehicles is seen to be even more acute.

The increasing flow of vehicles has by no means saturated the road. Applying the usual criterion of vehicles per unit length of road, the Arctic Highway will not see traffic jams, except perhaps within the largest two towns, for some years to come. However, the larger number of cars and, especially, the increasing use of the road by heavy goods vehicles, thus disproportionately raising the passenger-car-unit level, has placed a burden on a highway which was originally designed to carry fewer and lighter vehicles. The soft edges, gravel surfaces, narrow bridges and sharp bends ill-fit today's traffic.

The concentration of traffic along particular parts of the Highway is a further problem, for if special attention is given to these sections at the expense of the less well used segments, then the uneven patterns of traffic may well be perpetuated and the Highway never grow into the continuous link road for the arctic fylker which was the intention of its planners.

E

Nonetheless, to satisfy the demands of the most heavily used parts of the Arctic Highway, they have often been given priority in the programme of improvements to surfaces, road widening and straightening. Thus, all the urban areas along the Highway have asphalt surfaces and their approaches are mostly made of oil-bound gravel. This latter surface, which has the appearance and *feel* of asphalt, is probably the material of all future improvements in non-urban areas. Even at approximately 30,000Nkr per kilometre it is cheaper than asphalt and, more important, easier to repair. It is suitable for all but very heavy density traffic and will ultimately replace the cheaper but less suitable water-bound gravel.

There are problems in the selection of gravels to be mixed with the oils but, on the whole, these surfaces have stood the test of heavy traffic even after the difficult seasons of spring and winter. The climate is not so severe as to demand the use of less conventional materials such as polystyrene insulation board which are the subject of highway experiments in Alaska.

The longest single improved section of the Highway is across the Saltfjell, where it now is possible to drive from Mo i Rana to Fauske on surfaces which are almost one hundred per cent asphalt or oil-bound gravel. Water-bound gravel has also largely been eliminated from the Highway near Vollan and on several long sections of the road east of Skaidi and north of Banak. Nevertheless there is still much more re-surfacing to be carried out and, with limited funds, there will be rough gravel sections to the Highway for some years to come.

The annual programme of re-surfacing is approximately 250km of oil-bound gravel and 150km of asphalt, but this includes the E6 between Mo and Trondheim and, because of heavier traffic, much of the work is necessarily concentrated on these southern sections.

Concurrent with improvements to surfaces has been the construction work aimed at keeping the road open for longer periods during the year. Up to 1967-8 there were still some four lengthy sections subject to prolonged winter closure. The Highway across the Polar Circle, the Saltfjell, was usually closed from the end of December to the beginning of May. The Sennaland section, north of Alta, and Børselvfjellet, between Børselv and Storfjordbotn, were not open to traffic from mid-December to mid-May, and the Highway between Ifjord and Bielv was closed for fully half the year until early June. Since then two of these sections have been kept open during the winter season. Saltfjellet carried traffic first in the winter of 1968-9 and the long Sennaland path was free of winter closure only as recently as 1969-70. Of course, much depends on the severity of the winter and the Highway is kept open as long as ploughs can get through, but traffic must often be halted on various parts of the road for a matter of days or weeks if snowfalls are especially heavy. The declared aim of the *Vegdirektorat* is gradually to reduce closures to a minimum and even to look to a time when no section will be shut off to traffic for more than a few weeks. To achieve this, the post-war reconstruction programme has made an effort to raise parts of the Highway above the level of its surroundings, thus reducing the accumulation of snow. This *lifting* has been successful in a number of vidder sections, notably east of Skaidi and on the Kvænangsfjell.

When the road is raised it is generally also re-surfaced and strengthened, thus permitting a higher axle-weight tolerance which, in spring, can lead to the road being closed to general traffic for a shorter period. Other minor snow-protection methods have been discussed in Chapter 1, but it must be admitted that, without expenditure of sums vastly greater than those presently available, little can be done. Local people, knowing the problems only too well, are sceptical of promises of an all-weather, twelve-month Highway.

Although there is still much to be done, the quarter century since the end of the war has seen splendid improvements. The reduction in the number of ferry crossings to four and the generally improved engineering of the Highway are in the best traditions of the North Norwegians' fight for supremacy over their environment. However, like any other road, the Arctic Highway's story is never completed and the last thirty years of this century, to be discussed later, are likely to be as exciting as the previous half century during which the Highway has evolved.

Chapter Three

Mo i Rana to Narvik

Mo i Rana has many claims to be better known than it is. Even inside Norway, the arctic towns of Tromsø, Narvik and Hammerfest more readily come to mind. Yet Mo's population is not much less than Narvik's and nearly treble that of Hammerfest. Perhaps its isolation at the head of the Nord-Rana arm of the deeply penetrating Rana fjord has something to do with it; that and the fact that it is off the track of the *Hurtigrute*. The Rana fjord thrusts so far eastward into the uplands that Mo is only some fifteen miles from the Swedish border.

All the same the town is full of interest. Not only is Mo at the geographical centre of the country, but it can claim to be the first of the Norwegian arctic towns, nestling in the shadow of the great Svartisen plateau glacier, and the last of larger settlements on the land routes north to be encountered before crossing the Polar Circle. It is at Mo that the Arctic Highway begins its almost 1,000-mile journey to the far north. True the road is strictly a continuation of the E/R6, now much improved in its mountainous route from Mosjøen to Mo, but no road across the Polar Circle and into the Arctic could be so unromantically described as the E/R6—no, from here it becomes the *Arctic Highway*.

Although there were Viking settlements in the Rana fjord, no town of any consequence arose at the head until the mid-nineteenth century. The fjord is the natural outlet, not only for the farms of the broad Ranelv, but also for Överuman and Tärnaby in what is now Sweden. Nevertheless, even in the

MAP 2

eighteenth century, few had settled here. Ytteren, now a north-western *suburb* of Mo, had the densest population and here were sited the boathouses of the valley farmers. Hemnes, characteristically on an island, was the leading settlement of the fjord and the valley of Dunderslandsdalen remained only sparsely settled even in the early 1700s.

In October 1716 the *Lapp's Apostle*, Thomas von Westen, came to Mo i Rana on the first of his many missionary journeys to the north. He found a mission school had already been opened for the Lapp children of the area on Moholmen (now part of the waterfront). After several visits von Westen, who was also responsible for Trondheim's *Seminarum Lapponicum*, had raised, by 1722, enough money from the local farmers for a church to be built on the rising ground to the east of the fjord-head flats. Visible to travellers arriving at Mo by boat along the fjord, the church with its Dunderlandsdalen tarred exterior was a landmark and focal point for the Rana community. Mo's church today is on the same site, although it was entirely rebuilt in 1801 and now, unfortunately, is hidden by large modern buildings.

Slowly the settlement grew, becoming a parish independent of Hemnes in the mid-eighteenth century. But, fifty years later, when the combined population of the Hemnes and Mo parishes was 4561, the vast majority still lived in Hemnes. It took another sixty years for Mo really to be put on the map.

The making of Mo i Rana was the responsibility of one family, the Meyers. The family had traded in the area for decades, largely at Nesna at the mouth of the fjord. The son of the family, Lars Aagaard Meyer, moved to Mo from Vikholmen in 1860, when he was not yet thirty years old, and obtained a trading licence on 21 May of that year. Lars Meyer bought virtually the whole of the present town area which consisted then of little more than a few houses. From this time onwards the town has grown.

Meyer's skill as a merchant quickly became known. His principle was to act as a middle man. He found markets for the produce of the area and stored in his warehouses the goods needed by the local community. Trade flourished; everything Meyer organised seemed to succeed. A road was built between Mo and Sweden, to become known as the *butter road,* as Swedish produce came into the town. Many hundreds of boats were built from local timber to sail the length of Norway and across the world. When Lars' son, Hans, took over in 1902 Mo was already known as Meyer's Town, and even today the family still owns some sixty per cent of the town area.

Mo i Rana became separated from the Nord-Rana Kommune in 1923 and plans were made for a town of 5,000. However, the difficult world economic climate of the early 1930s, coupled with local unemployment caused by the decline in iron ore mining, halted growth. The population failed to reach 1,500 by 1930, although it had been 1,457 in 1923, and only in 1935 did it just exceed 1,800. The late 1930s saw greater prosperity and growth with the building of the Nordland railway and the Arctic Highway across Saltfjell, but it was not until after the war that the second landmark in Mo's history—comparable with Lars Meyer's coming—occurred.

Partly through lessons learnt from the war, the Norwegian government was determined to strengthen the economy of North Norway: on 10 July 1946 the Storting unanimously passed a bill to set up a national iron works (*A/S Norsk Jernverk*) in Mo i Rana. This act foreshadowed the North Norway Plan published some five years later, setting out the socio-economic aims for the region in detail.

By the mid-50s the steel plant was in production and a decade later had doubled its output. Today it is the largest steel plant in Norway, aiming to produce half Norway's requirements and, since the mid-60s, using local ores as well as imports from Kirkenes. More than half the town's working population is employed in the steel works.

From this Norwegian *Sheffield,* the Arctic Highway starts its journey north, first as Sørlandsvegen, then as Nordlandsvegen. In fact little of the jumbled town can be seen from the Highway. Only when flying over it can its patchwork development be appreciated. On the ground Mo appears bustling and work-a-day but undistinguished. The hub of social activity is around the Railway Station Square. Here is the most helpful tourist kiosk I have ever encountered, facing a well-kept garden tended by the ladies of the town. Besides the railway station and a taxi rank, there is a large store and, across the square to the north, the Meyergarden Hotel which was the home of the Meyer family. The hotel is a good example of the splendid houses typical of the nineteenth-century merchants of the area. The painted ceilings, well-furnished public rooms and good food make this one of the best hotels in North Norway. Apart from the hotel, two other buildings are worth visiting: the museum and, on the other side of the Highway, the church.

The church of 1801 was altered in 1832 and in 1860 to its present form with high gabled roof and onion-spired low tower. Typically simple in design, the wooden exterior is now gleaming white, having been bright red until 1908. The interior has some most interesting relics from the first church—none more so than a pauper's purse and poor box—but also later additions. The painted altar piece is very fine, but even more fascinating, in a chapel behind the choir, is a simple carved altar-piece dating from 1768. As so often in North Norway, some of the wooden interior is painted to resemble marble.

But Mo is more than museums and churches, it is a thriving industrial settlement of some 10,000 people. Driving along the Highway through the town there is everywhere the sign of industry, not least in the pink tinge given to buildings, pavements, even to gardens and lamp posts, by the outpourings of the steel work's chimneys. The unsuccess of Norwegian and British scientists who have tried to reduce the pollution is everywhere evident. The steel works dominates the town, occupying about half its area with giant rolling mills (the largest covered building in Norway), overhead cableways and shutes, and fuel stores. Meyer's Town has become Steel Town.

Plans have been mooted for the Arctic Highway to bypass Mo i Rana, but this, in my view, is as unlikely as it is unnecessary. True, an improved link with the road to Sweden, Route 77, must be made, but this latter road has already been straightened and widened within Mo and improvements to the mountain section agreed with the Swedish authorities interested in exporting timber through Mo. At present the road up to the Swedish border is a nightmare of narrow and steep hairpin bends on a gravel surface. (The police chief at Mo once told me of a British tourist who became so scared when descending this mountain road that he abandoned his car and refused to continue his journey into Mo!) When the standard of Route 77—the old *butter road*—is raised and the link made with the Arctic Highway, it will greatly improve traffic flow into the town.

The Arctic Highway leaves Mo i Rana, passing some new multi-storeyed flats, dipping under an overhead conveyor to the steel works, past a jumble of small factories and crossing the Rana by the post-war Selfors bridge. Here, in summer, small boys will stop foreign cars, asking for autographs and coins.

East of the bridge is the *suburb* of Selfors, a sort of small factory estate, but gradually the Highway leaves behind the scattered outskirts of Mo and threads its way along the right bank of Ranelva. Soon the asphalt gives way to a gravel surface, but one scarcely notices the change as the magnificent river gorge comes into view. Deep below the Highway the broad Rana, turbulent and turquoise, separates the road on the right bank from the railway on its southern side. The river, in the summer sun, reflects the blue-black faces of the gorge in its glacial melt water.

Both the Highway and the railway are engineered out of the sides of the gorge. The railway makes use of tunnels to reduce its gradients, but the road clings to the canyon's walls. The railway, now part of the Nordlandsbanen line through to Bodø, was a British contribution to communications.

In the 1870s Ole Tobias Olsen and Consul Nils Persson competed for mineral rights in the Dunderland valley after Swedish geologists had established that iron was probably present in significant quantities. Persson eventually gained control and, with the help of a Professor Hasselbom, proved the deposits sufficiently rich to attract the interest of British industry. In 1905 the

British Dunderland Iron Ore Company moved in and, to enable the export of ore, built a standard gauge rail line along the seventeen miles which separate Mo from the mines at Storforshei. The mining was not entirely successful, for the ore is phosphoric and not easily worked. Although the company lost considerable sums of money, it remained in operation until the outbreak of World War II, by which time the railway was destined to become part of the national system.

While the railway stays with the river, the Highway moves away to the north through a profusion of rock outcrops and glacial deposits. The land is partly farmed and partly forested with mixed conifers and birch. At Reinforshei the Highway crosses the lactescent Blakkå, tributary of the Rana, which is part of the Svartisen glacier's outflow. Nearby, the Reinfors salmon ladder climbs round a hydro-electric station's dam. The ladder follows a tunnelled route and at 370 metres claims to be Europe's longest. Two miles further on is the village of Røsvoll. Here an octagonal church, built in 1953, stands by the Highway. It has a concrete base but is otherwise wooden, and the light pastel shades in which its interior is painted are no less attractive because they are typical. In winter the church is floodlit. Close to the church a rough gravel road leads off to Svartisen and the Grønli caves (see Chapter 6) and to the new airport.

Until 1968, Mo's airport was simply a landing strip cut out of the forest for the local flying club. The Mo Flyklubb was founded in 1959 by a group of non-professional flying enthusiasts with a Piper seaplane. In 1963 they started to clear a rough landing strip in the forest near Røsvoll and a year later the first plane landed. In 1968, on 1 July, the landing strip became Mo's airport, with a new hard runway and a control building. The State and the town shared in the purchase in anticipation of Widerøe's new link with SAS which serves the smaller towns.

The flying club (see also Chapter 8) has benefited from the purchase and bought a new plane out of the proceeds. It continues to use the runway and also has seaplane moorings on Langvatnet.

Beyond Røsvoll the Highway has an excellent oil-bound gravel surface to carry it through Dunderlandsdalen and across Saltfjellet and the Polar Circle. The Dunderland valley is one of the oldest settlement areas in the Rana region: a mixture of old farms with their turf roofs and weathered wood exteriors, new brightly painted farmhouses and the occasional nucleated settlement as at Storforshei and Dunderland. The Storforshei community has grown since the working of new iron ore deposits from 1964. Most of the buildings are undistinguished, but the settlement is well served, with a small shopping centre and a modern school. Like so many mining villages it has that unfinished, pioneer appearance.

Dunderland is a railway village with sidings and maintenance sheds. It lies just off the Highway and is the depot from which the whole of the railway line from Mosjøen to Bodø is serviced. Great rotary snow ploughs are kept here which generally manage to keep the line open throughout the winter months. This is an especially isolated and somehow sad little village with few amenities.

The landscape along this part of the Highway is characterised by high mountains enclosing the generally broad Rana valley floor. The river has cut deep terraces in the glacial deposits and the road as well as the farms make use of these flat ledges. In some parts the river valley narrows, farms disappear and railway and road are again forced into close proximity with the river. Such is the situation in the Illhullet gorge where a ledge on the right bank carries only telegraph wires where it once formed the path for the railway. Now the line burrows its way by tunnels through the mountains. Other tunnels, visible from the Highway, allow Svartisen to discharge its potentially dangerous melt water into the Rana.

Ørtfjellet and Jarfjellet rise to around 4,000ft and are capped with snow patches even in mid-summer. The lowest slopes are forested, although only occasionally, as at Storforshei, is there much exploitation, commercially, of the timber. The road rises and falls as it crosses hummocky morainic deposits. At intervals narrow tracks lead off the Arctic Highway to farms, some of which are grouped into small hamlets like Eitevåga. Both sides of the valley are farmed with small suspension bridges giving links between the left bank and the Highway.

As along the whole of the Highway, the farming is nearly exclusively cattle reared for dairy produce. Hay is almost the only crop, providing as it does the necessary winter feed when, in the long winter months, the valley is choked with snow. In summer cattle roam along the roadside and in the woodland while dun-coloured horses work the fields and forest.

The Rana is crossed by the Highway on a steel girder bridge at Messingslett. The Germans destroyed the original stone arch bridge and temporarily replaced it with a wooden structure. Once over the bridge, the road continues on to the north-eastern end of Dunderlandsdalen by the left bank of the river. For a while it is alongside the railway some 200ft above the river but descends again to the confluence of two glacial valleys at Storvollen. It is here that Tespdalen and Dunderlandsdalen meet and the terrace platforms are heavily farmed. In the village is a monument with a bust of Ole Tobias Olsen.

Olsen, who is described as *father* of Nordland, was a parish priest and teacher passionately interested in the development of North Norway. After a visit to the 1862 London Exhibition he conceived the idea of a great international railway through northern Europe. He had seen fish arriving by rail in London from Scotland and he foresaw a time when Norway could export fish to Russia in like manner. In 1872, when he was also concerned with the exploiting of the Dunderland iron ore, he wrote of a rail link between North Norway and St Petersburg via Happaranda on the Gulf of Bothnia. The idea met with support from many quarters. *The Times* went so far as to suggest an even mightier scheme, with a continuous north Eurasian line to Peking and branch lines to India. Olsen produced a blueprint for his railway in 1874, yet it took fifty years of campaigning before the Storting, in 1923, agreed that a North Norway railway should be built. A year later, aged 94, Ole Tobias Olsen died and some twenty years later the railway passed through Storvoll on its way across the Polar Circle.

Outside Storvollen the road passes under the railway (*see plate, page 51*) and climbs the wooded Randalselv. The valley sides are steep and gullied by streams pouring from Saratuva and Kjerring-fjellet, mountains exceeding 3,500ft. The few farms along the valley are frequently isolated for weeks during winter but in summer huge piles of logs only hint at the severity of the months to come.

Randalselva is little different from Dunderlandsdalen, but beyond Krokstrand the whole character of the road and its surroundings begins to change. It is not that the road is wider or narrower, straighter or more winding; even the surface continues to be excellent oiled gravel. But from Krokstrand the Highway starts to climb to the top of the vidda, to the Polar Circle. The Randalselv is crossed above the Silfoss waterfall and the Arctic Highway strikes north, first along a section of Roman straightness, then winding its way speedily upwards, breaking out of the confines of the valley into more open country. The forest becomes thinner and the trees stunted; birch replaces pine and, by 1,500ft, the country on either side of the road is strewn with boulders, glacial erratics. The Highway rides above the valley on the right-hand side and countless incised tributary streams are crossed by narrow four-metre stone bridges.

This part of the Highway is patrolled by the NAF (the Norwegian automobile association: *Norges Automobil-Forbund*) across the Polar Circle and the Saltfjell during the summer. One patrolman in his own car is there to aid motorists in trouble or assist at the not too infrequent collisions that occur on the narrow gully bridges. The occasional NAF telephone is to be seen tied to a telegraph pole but the volume of traffic does not keep the patrolman especially busy.

A reindeer warning sign correctly suggests Lapp encampments, and mid-summer snow patches are a further reminder that this is the arctic. By 2,000ft the trees have been left behind, the road is on the vidda and the landscape has taken on a strange lunar quality. Erratics and low gravel terraces are evidence of glacial and river morphology. Much of the surface is covered with a blanket of peat bog on which some vegetation rapidly returns to life when the snow melts each May.

The uninhabited Saltfjell calls for no branch roads from the Highway. Vehicles race across the barren wasteland eager to return to the more sheltered valleys to the north or south. Few tourists, however, will forgo a stop at the Circle itself. It is marked by a signpost, by a stone monument and by cairns topped by open metal lattice globes. Since 1956 there has also been a hutted café (*see plate, page 52*) at the Circle. Previously, from the time the road was opened in 1937, souvenir sellers sat out in the open during the daylight hours of the tourist season. The café is staffed from the nearby Høyfjellshotell and remains open from the end of May until mid-September. The café is unashamedly tourist. Here are the certificates to prove a crossing of the Arctic Circle signed by *Jack Frost* or by the *King of the Glaciers*. Postcards can be bought and given a special Polar Circle stamp and all manner of souvenirs are available, from imitation Lapp clothes to those gaudy knick-knacks which seem to turn up in tourist kiosks the world over. Language is no problem; the café staff are mostly students with the gift of tongues.

Just south of the Circle is a monument to the Yugoslav prisoners who worked on the Highway during the war (see Chapter 2). The inscription, in Norwegian and Serbo-Croat, is simple: *In memory of the Yugoslav partisans and political prisoners who died as a result of Nazi terror in Norway : 1942-5.* It is saddening to find that German tourists have defaced part of the monument. North of the Circle the remains of the prisoners' hutted camp, with its concrete floored cookhouse, can be seen.

The prisoners failed to keep the road open through the winter and the people of this part of Nordland take it for granted that short-period closures will continue to be necessary. Annual promises by the Highway Authorities that the next winter the road would be cleared of snow were treated with a scepticism born of experience. Then in 1968-9 the road *was* kept open. The road is really in the wrong place. Had the builders heeded the advice of the Swedish Lapp family Blind, who graze reindeer on this plateau, the Highway would have been built away from the slope down which the snow is blown to collect on the largely unembanked road. Too late for a complete re-routeing, the Highway has been raised to produce snow traps on either side and this, and the use of blower ploughs, keep the road open almost all the year.

In contrast, the railway, on more open ground and protected by wooden tunnels (*see plate, page 52*), is relatively easily kept open with the help of rotary ploughs from Dunderland. Cars can be ferried by rail across the Polar Circle even in mid-winter.

The watershed of the plateau is crossed near Stødi and the road reaches an altitude of 2,300ft. Northwards the vidda continues and the Highway crosses over the railway (*see plate, page 52*), where barriers are put across the road during winter or spring closures. Now Lønsdalen is the Highway's route line and, by 2,100ft, the forest begins to return and the road descends alongside Lønselva. The river tumbles over a series of shallow steps, trees fill the valley, but bare rock outcrops on the exposed higher slopes. The railway, high above the road and river to the west, is the life-line of the Arctic Circle Hotel, the Polarsirklen Høyfjellshotell. The new post-war building lies just off the Highway where the road descends into Lønsdalen. It is open most of the year, but winter is the busy season, for it serves the skiers who use the magnificent slopes of Saltfjellet. Access in winter is by way of the railway, although summer visitors are usually motorists. A sauna bath and well-furnished public rooms allow one to forget the rigours of winter, while the views, especially eastward towards the Swedish border, are breathtaking in the soft light of an arctic sun.

As the Highway sweeps down Lønsdalen, it is confronted by the bleak exposed face of Solvågtind to the north of Junkerdalen but, as the road descends, the valley narrows and the peaks no longer capture the attention. Conifers replace birch at 500ft and the valley becomes choked with trees, while above there is only

Page 87 Isolated northern farm, the home of a fisherman-farmer near Alta. The Highway runs on the terrace to be seen on the right of the photograph.

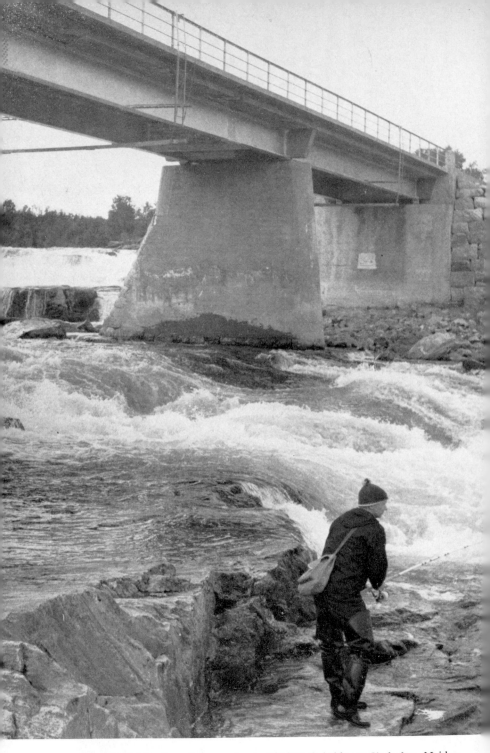

Page 88 Fishing for salmon under the Highway's bridge at Skoltefoss, Neiden. One of the many good northern salmon rivers, Neidenelva is less well known than the Altaelv or Lakselva.

bare rock. Even in high summer the mountains retain their snow, which spills off the summits and into the gullies. Beyond the confluence with Junkerdalen, the valley, now called Saltdalen, remains narrow but widens after Bleiknesmo (*see plate, page 69*).

Much of the floor is cleared of forest and the lower valley terraces are farmed. Ground by the river is boggy in summer, but the stream is rich in salmon. The farms tend to be isolated, one from another, each occupying a suitable patch of dry terrace. Their prosperity often can be judged by the state of the paint on their wooden buildings, but most are well kept and, more often than not, will be flying the Norwegian flag on a high pole throughout summer. In winter the isolation is harder to bear, for there are few settlements which could be ranked even as villages.

Closer to Saltdalsfjorden the valley widens appreciably. The Highway crosses the river by a two-span girder bridge and the railway is crossed at a cluster of little hamlets which include Røkland, Vensmoen and Kvæle. This part of Saltdalen has been settled for centuries and was formerly important for boat building. Today the settlement is still sparse, although increasing in density as the Highway passes through Medby and Nestby and into the fjord-head village of Rognan. At Medby there is an improved mountain road to Skjerstad. This may become the Highway's link with the proposed coast road (see Chapter 8).

Rognan has a population of just over 1,000 and used to be the ferry point for Langset (see Chapter 2). Now it is just another Highway village with unmade roads and typical pioneer appearance. Its church, built in 1862, is of interest but the village is otherwise undistinguished. A fibre-board industry and a little boat building occupy its inhabitants.

Out of Rognan the Highway turns eastward then northward to follow the edge of the Saltdal arm of Skjerstadfjorden. A fine piece of engineering carries it on a cliff ledge to Botn, while the railway tunnels through promontories.

At Botn the railway runs parallel to the Highway and a half hidden signpost points towards a cemetery. This is no ordinary graveyard. Half a mile on the other side of the railway, up a steep

F

slope, is the resting place of over 4,000 bodies, none Norwegian. Here are buried 1,657 Yugoslavs and 2,732 Germans from the last war. National differences are carried to the grave, for there are two separate cemeteries. The Yugoslavian section lacks the care and attention obviously given to the German part. The latter is carefully tended, the graves numbered and named, and in the entrance lodge a book records the many visits of relatives who not infrequently make written corrections to the list of those buried. Among the Yugoslavs buried here is the unnamed prisoner who was the first victim of Nazi tyranny in Saltfjorden. His *crime* was to open a door in the prison before the appointed time of ten o'clock in the morning. He was shot.

The terrain of the fjord coast is difficult and the Highway's path towards Fauske is often on ledges blasted out of the rock faces above the water's edge. Protection from rock and snow avalanches is given by concrete canopies up to 100m long. Crash barriers have been built and two tunnels now allow the road to escape from the narrowest ledges. The newest tunnel, about half a mile in length, takes the Highway below the old road but it climbs again from the fjord edge to cut off Leivset peninsula and enter Finneid.

At Finneid the railway, which has reached here largely through tunnels, is met by the Sulitjelma copper mine line. Railway and Highway share a bridge into this settlement, which is contiguous with the small town of Fauske, but part company beyond the town. Westward the railway continues to Bodø, leaving the Arctic Highway to follow its lonely journey to the north.

Fauske is very much a route centre. The Highway and the Nordland railway pass through and diverge; Route 80 makes a road link along the fjord to the Nordland capital of Bodø; and from the east comes the little mountain railway which is Sulitjelma's only link with the outside world. In the shadow of the Blåmannsisen glacier, pyrites and copper concentrates are mined to be sent by the railway for export from Finneid. If ever a settlement merited the expression *God-forsaken* it is Sulitjelma and its neighbour Jakobsbakken. These mining villages share with a few farmsteads almost complete isolation in the mountains above Fauske. Yet if one ignores the dust showered on the countryside by the great electric smelters, if one is content to be cut off from the turmoil of the twentieth century, then here is a haven. Magnificent in a white arctic blanket for so much of the year, the mountains become a patchwork of rock, snow and bright alpine flowers in the few weeks that are summer. Even the glacier, bared of its snow cover, glistens and the isolation can be forgotten or enjoyed.

Down by the fjord, Fauske shows no signs of the almost total destruction which resulted from a bombing raid in May 1940. The town has a number of important buildings including a high school, a mineral waters factory and municipal offices. There is nothing very grand about Fauske but this is a busy, purposeful community.

On one side of Fauske is the fjord and the jetty from which marble, quarried and polished locally, is exported. On the other side, the northern perimeter, is the bus station, an important place, for Fauske is also the starting point for the North Norway Bus. This scheduled bus service, which uses exclusively the Arctic Highway, was begun in 1947 to give a complementary service to the Nordland Railway. Although a number of independent companies are involved, the service allows passengers to embark at Fauske and be carried in comfortable coaches to Kirkenes at the northern end of the Highway. Three overnight stops are necessary—at Narvik, Sørkjosen and Laksclv—but the bus is a splendid way to see the country and its people for, to those living along the road, it is their local bus. The service from Alta to Kirkenes is usually stopped from the beginning of October to the beginning of June and occasionally other sections are curtailed by snow or spring melt. For the rest of the year the yellow and red Nord-Norge-Bussen is the world's only true arctic bus service (*see plate, page 33*).

Several roads, including the Highway, cut across the Fauskeidet north of the town. The route is easy across the neck for the valley floor is wide, but marsh and poor drainage make the gravel surfaces of the Highway very soft in spring. The low relief allows farming and forestry to take place and the local quarries can be seen on the western side of the road. To the east rises the conical mountain Rishaugfjellet. Its steep bare sides are often clear of snow even in winter, but in summer a banner cloud is usually attached to the summit and the mountain looks like a smoking volcano.

After five miles the Highway curves eastward, leaving behind its old route to Røsvik, the former ferry point to Bonnåsjøen. This is one of the newest sections of the Highway (see Chapter 2) and a fine asphalt surface contrasts with the preceding water-bound gravel. After the junction with the road to Grønnås, the Highway rises gently above the valley floor allowing the river to broaden and flow into Sørfoldafjorden.

Near the mouth of the fjord is yet another of those arctic surprises. In an otherwise totally rural setting, twin factory chimneys suddenly break the horizon, belching white smoke into the clean air. This is the new Valljord ferro-silicates works sited at the fjord-head where raw materials can be brought in by ship and where the mountain streams provide hydro-electric power. The labour comes from Fauske. Fortunately the factory is dwarfed by the scale of the landscape and conservationists rest easy.

The Highway reaches the fjord by way of a curving path, rising at first then falling towards the sea. Much of the road has had to be blown out of the rock face but it is wide enough. To the north is the fjord and to the west mountains rise, their slopes covered with snow in winter but streaked with silvery waterfalls in summer.

The next twenty or so miles of the Highway are the pride and joy of Nordland, a triumph of man over nature. As described in the previous chapter, this section has been open only a few years and has grossly reduced the ferry link across Sørfolda. The engineering is especially fine and this whole twenty-four-mile section is probably the Highway's best. Something of the complex nature of this part of the road can be seen in the diagram opposite.

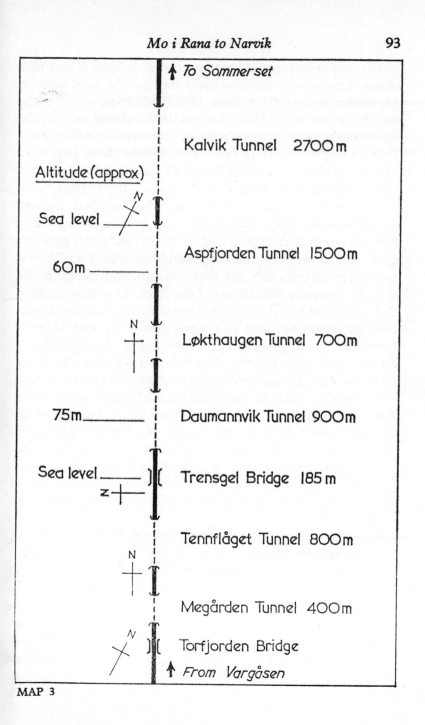

↑ *To Sommerset*

Kalvik Tunnel 2700 m

Altitude (approx)

Sea level

Aspfjorden Tunnel 1500 m

60 m

Løkthaugen Tunnel 700 m

75 m

Daumannvik Tunnel 900 m

Sea level

Trensgel Bridge 185 m

Tennflåget Tunnel 800 m

Megården Tunnel 400 m

Torfjorden Bridge

↑ *From Vargåsen*

MAP 3

The road skirts round the eastern edge of the fjord at first, then climbs a ridge only to drop again to cross the fjord-head by Torfjorden bridge. Again along the water's edge, the Highway threads its way to the Megården tunnel. Absolutely straight, the quarter-mile tunnel is electrically lit but eerie: one imagines a Troll lurking behind every rocky outcrop. Like all the Highway's tunnels, this is blasted out of the solid rock and unlined. Emerging out of the northern end of the tunnel, the Highway runs round a narrow branch of Torfjorden, the Nordfjord, through the half-mile Tennflåget tunnel and under a further short twenty-five-metre rock arch. The Trengsel bridge (*see plate, page 34*) spans the six-hundred-foot-wide Nordfjord with graceful ease and on the northern side the Highway falls, then climbs gently to 250ft above the dark waters of the fjord. Two more tunnels, each about half a mile in length, keep the road a couple of hundred feet above sea level and carry it over the spur between Torfjorden and Aspfjorden. Below, the hummocky glacial headland is farmed from Kvarv and on the other side of the Highway mountain lakes rest like jewels enclosed by tree covered slopes.

For nearly a mile the Aspfjorden tunnel winds its way back down to sea level and takes the Highway into the longest of all its tunnels, the Kalvik (*see plate, page 34*). Running parallel to the Leirfjord arm this tunnel is straight, level and with lay-bys in case of breakdowns. Finally the Highway breaks out into the open and, with the exception of a short rock arch, remains so until the Sommerset ferry point is reached halfway along the fjord.

In summer, when traffic is heavy, two ferry ships, the *Rombakfjord* and the *Røtinn,* make the short journey down into Leirfjorden and across to Bonnåsjøen. No time is wasted. Vehicles follow passengers off the ship as soon as it is secured at the jetty and immediately a fresh human and wheeled cargo takes its place for the return journey. There is scarcely time for tickets to be issued or the saloon café to be visited before, with a slight shudder, the ship announces that it has crossed the fjord and is ready to disgorge its cargo again.

In fact, the crossing, like those of all Norwegian fjords, is deceptively simple and the skill of the half dozen crew disguises the treachery of the inlet's currents. Leirfjorden has, indeed, a somewhat awful appearance. Clothed in a mantle of snow and cloud, Rismålsfjellet and Grønfjellet to the west and Sommersettinden to the east so tower above the fjord that the impression is almost that of being at the bottom of a well.

Surprisingly, some small farms are scattered along the north facing edge of the fjord and crowd its head. Fishing and farming combine to provide a livelihood but isolation is almost complete. There is no continuous road link with the Highway and the communities use small rowing boats to take them to the ferry points.

Bonnåsjøen is at the only possible place for a road on the north side of the fjord. Until one actually arrives at the jetty it seems impossible that the Highway could have any route at all through the mountains. But here there is just enough room for the jetty, a café and a petrol store. The Highway climbs steeply away from the water's edge through the narrow and winding Bonnelv valley. In five miles 300ft is reached. Birch trees cling to the rocky valley sides and turbulent streams tumble over falls in their eager rush to the sea. This is a very difficult section of road in winter, for the very narrowness of the valley ensures that it will act as a snow trap. In summer, the steepness and gravel surface can be forgotten in one's delight at the scenery.

As almost everywhere along the Highway, the scenery has three basic elements—mountain, water and sky—yet they are never the same. Changing in mood with the time of year or time of day, at once forbidding and inviting exploration, the landscape has a grandeur all its own. Happily man's contribution to the scene is rarely out of harmony with Nature. In any case, competition cannot be on Nature's scale.

The tiny but charming hamlet of Bonnå occupies a site where the valley widens and Horndal is built on the shore of the lake which shares its name. Both hamlets are overshadowed by Blåfjellet and Horndalfjellet, whose bare slopes contrast with the birch filled gullies and lower valley floor. Although the Highway rises only to just over 500ft, there is little vegetation except where shelter is afforded and soil has accumulated. The mountain slopes are foliated and weathered scree slopes rest on ledges or valley floors. As in Leirfjorden, there is a sense of enclosure.

The Highway drops to Mørsvikfjorden, an arm of Nordfolda, at the head of which lies Storeidet, built on delta flats and terraces and indistinguishable from its neighbours Sildhopen and Mørsvikbotn on the other side of the inlet. These are not villages but a handful of little houses, brightly painted yet somehow part of the natural scene.

The Highway runs round the head alongside massive screes which spill off Sildhopfjellet. Impressive waterfalls descend into the Mørsvikfjord whose steep sides are unsettled. Quickly away from the fjord, the road again ascends up the deep and narrow Mørsvik valley to Mørsvikvatnet. From here water seems as though it may dominate the scene.

Lakes, some with narrow white beaches of quartz sand, appear to lie in every hollow, at every altitude. Waterfalls crash over precipitous slopes or fall gently over natural steps, reminding one of the gardens of Chatsworth House. Many of the falls make use of the simple rock structures, racing over bedding planes and falling over the edges. Where there is shelter, birch trees have again colonised. If the ground is flat, a thick fern carpet grows on the moist peaty forest floor in summer.

But always the mountains eclipse the rest of the scene and it is the mountains which force the Highway to pick its path over shallow cols and along knife-cut valleys. The road must rise to nearly 1,300 feet to get through this strange uninhabited region and to Kråkmoen.

Kråkmo is notable as a base for Nordland's *Statens Vegvesen*, whose large sheds house a collection of snow ploughs and road building equipment. Here too is Kråkmotind, a near-3,000-foot mountain peak. But peak is the wrong description, for Kråkmotind's top appears to have been sliced off to give the mountain a peculiar truncated, sugar-loaf appearance. Most of the mountain summits are permanently above the snow line which slips well down the free face on the north facing sides. Lower down are scree girdles and below that the mountains are encircled by forest.

Many of the old bridges across mountain streams have been replaced and the road re-aligned, leaving behind the old abutments as evidence of the previous path. In the stream beds quartz veins glitter through the clear water. The abundance of water has not been ignored and hydro-electricity is generated at a station just north of the village. Huge pipes stretch upwards and out of sight over the mountain shoulder.

Towards Sagfjorden the Arctic Highway runs alongside a series of large lakes. Altogether there are seven, the first three named, the rest numbered. Each lake drains into the next over low moraines, sometimes by way of shallow rapids. The lakes are enclosed by high dome-shaped mountains of bare rock rising to over 3,000ft. There is little room for the road, which has to make do with a ledge above the lakes and only occasionally is there space for more than an isolated farm. At Sandnes a cluster of some half dozen or so houses nestles by the lakeside. Tall flagstaffs proudly flying the national flag stand in neat little gardens and the influence of tourism is seen in the small café and a camping site.

Eventually the Highway reaches the open fjord at Vassmo, where rock carvings claim settlement from the Stone Age, while a little further on, at Tømmernes, is another more recent but equally interesting monument. The hamlet consists of little more than a church and a few houses lying back from the road, but to the west of the Highway is a graveyard and a nearby monument inscribed: *To the memory of the Soviet soldiers who gave their lives in Norway 1941-5 and who are buried here.* One is reminded again of the Highway's other name: *The Blood Road.*

The road leaves the fjord at Innhavet. This rather untidy village has a small hotel or *gjestgiveri* but there is little to detain the traveller. The only attractive features are some massive ice-scarred roches moutonnées to the east of the road and a view across the high-walled Sagfjord towards islands and skerries which fringe the coast. At sunset or in the low winter sun the islands have a special beauty.

Northward the Highway strikes along the peninsula between Sagfjorden and the Tysfjord. The road surface is still rather poor gravel and widening is needed. Much of the immediate surroundings is bare rock; elsewhere, on thin soils, trees and bog occupy the lowest land. In contrast to the previous forty or fifty miles to the south this part of the Highway is without distinction. Occasionally the waters of the Innhavet inlet can be glimpsed and at Notvann an old water mill stands by the river crossing.

The neck of the peninsula is at Sommerset, where only three miles of land separate the Innhavet inlet from Tysfjorden. Continuing north, the Highway remains only a little above sea level until at Skilvassbakk it begins to climb again, following a winding path to 675ft. Descent from this high ridge is even more convoluted, the road in parts being dangerously narrow and in winter treacherous. The difficult route, however, is not without its reward. From here can be had the first clear views of the broad Vestfjord and of the Lofoten Islands which lie beyond.

At the bottom of the slope is Ulsvåg and the Highway is joined by the road (Route 81) which was built after the war to serve Hamarøy. This road also provides the Highway with a link to the intra-coastal ferries and an alternative route to Røsvik and Fauske from Skutvik. The village of Ulsvåg has a *gjestgiveri* and, as the largest community on the peninsula, serves the small farms which border the Highway along the edge of Stordjupet and around Bessfjorden.

The route from Ulsvåg is more or less direct to Bognes. First along Stordjupet to the old trading post of Sorkil, from where Tilthornet can be seen rising to 2,800ft, and on through a narrow forested depression, the road rises to 500ft. Down to Botn with views of the mountains to the north east of Tysfjorden, round Bessfjorden with its tiny summer houses and where the Highway is both narrow and poorly surfaced, and finally over a 250-foot headland and the road enters Bognes.

The last 1.7km of Highway to Bognes has a new asphalt surface and the ferry point is well constructed with adequate parking space for vehicles. The original ferry link to Skarberget for the Arctic Highway was from Korsnes—a few miles further north—but the new pierhead is a considerable improvement. Apart from the usual oil storage tanks there are no buildings except a café and a kiosk. In the café the waiting ferry passenger can have unlimited coffee for the price of a single cup.

Two ferry routes operate from Bognes. From the more northerly pierhead a ferry leaves for the hour-and-a-half journey across Vestfjorden to Lødingen. Once on Hinnøya, the largest of the Lofoten islands, Harstad can be reached by road on Route 19. The other pierhead is used to accommodate the motor vessels *Oskarsborg* and *Røffjell* which cross the five-mile mouth of Tysfjorden to Skarberget in just over half an hour.

The crossing is the most splendid of all the Highway's ferry journeys and is also likely to be the last to be replaced (see Chapter 8). On all sides of Skrovkjosen and on the east of the Tysfjord, the mountains rise vertically out of the sea, with Tepkiltinden and Breiskartinden dominating the north shore of the inlet. Just west of Skarberget great clefts in the mountain leave one with the impression that some giant has been busy with an axe. To the south east is the famous 4,500-foot Stetind, a pyramid of rock columns with claims to be the world's finest natural obelisk. In winter, from the air, the peaks and cols look like icing on a blue-black Christmas cake.

Skarberget itself has no more than a café to greet passengers off the ferry and the Arctic Highway slips away from the pier along the undulating shore of Skrovkjosen. The road surface is rather poor and the edges are soft, but the views are captivating. From Ulvik the little village of Hestnes appears across the fjord sheltered in a cusp-shaped embayment. All around are the ice-sculptured concave slopes of the mountains. The Highway has to rise to 800ft in crossing the peninsula between Tysfjorden and Efjorden. This section is fraught with difficulties in winter. The surface is poor and the road narrow, but snow ploughs stationed at the crest reduce the hazard.

West of the road is the conical Eidetind and to the east the smooth bare slopes of Huglhornet. The lower ground is stripped of all but loose rocks with only an occasional stunted pine to give a sign of life. As the road falls to Sætran there are glimpses of the narrow Efjord and of large roches mountonnées which border its shore. North of the fjord impressive peaks rise to over 2,000ft.

Sætran was previously the ferry point for Forså, but now three bridges use islands as stepping stones to carry the Highway over the fjord (see Chapter 2). The motor vessels *Solfrid* and *Frydenlund* have served their time. The new Highway is built above the water on the north side of the fjord. It has a very good surface, is almost perfectly straight and, what is more, eliminates the time-consuming ferry. Sætran had not even a café at its pierhead, so only the ferry company are losers.

The new section of Highway meets the old about a mile from the Forså ferry point with its rickety wooden ramp and piers. The population was increased temporarily by workmen living in huts while they built the tunnel which carries the road through the last part of its fjord journey. Beyond the tunnel much improvement has been made and further is planned to widen, straighten and re-surface the road.

It is from here that the influence of Narvik, still some thirty-five miles away, begins to be seen. From here and along the shores of Ofotfjorden and its inlets, the settlement is thicker and activity increases. At first the signs are inconclusive: goats and cattle grazing in thin forest, farmsteads a little closer together, concrete blocks replacing wood for some buildings. By the time the road enters Ballangen there is clear evidence of urbanism not seen since the Highway passed through Fauske. True, Ballangen has only a few hundred inhabitants, but it is industrial. It was once the port for the Bjørkåsen pyrite mines, which have long since closed, but today it is a rather shabby settlement—half town, half village—with one or two industries and completely without a recognisable form. It is scruffy in a way that characterises so many arctic towns. Land is cheap and the settlement is pioneer.

The Highway uses the same fjord shelf as countless small farms as it follows the shoreline of Ofotfjorden. This large open fjord contrasts with those typical of south-west Norway or even those encountered so far on the Highway. It is comparable, however, with those found in Finnmark. Where the shelf narrows, the line of farms breaks off, leaving the road to its solitary self. Vidrek is well farmed, but as the Highway enters the narrow Skjomen fjord, enclosed by steep mountains topped by snow even in high summer, the road is on its own again.

This arm of Ofotfjorden narrows at Grindjord. A short, ten-minute ferry linked the two sides, but now a new bridge will soon be opened just upstream from the ferry (see Chapter 8). The bridge is carefully sited where two promontories face one another across the water. On both sides a short rising section of road links the old ferry points with the new bridge which crosses from headland to headland above the fjord.

The path of the Highway into Narvik is again on a narrow shelf by the fjord edge. Thrusting their rusted frames out of the fjord, sunken ships remain as a reminder of Narvik's wartime history. The roadside buildings are now not all farms but commuters' houses. Although chiefly wooden, some are concrete and a very few newly built of brick. Greenhouses suggest Narvik's market influence and the asphalt surface of the Highway confirms the heavier traffic one has already noticed.

Suburban Narvik is reached at Ankenesstrand, which has a new octagonal church in wood. Neat houses, painted in pastel shades, line the slopes bordering the Highway: two or three rows on parallel un-made roads. A swing bridge crosses Beisfjorden and industrial Narvik is reached.

Narvik is the largest town on the Arctic Highway. Some would say it is the only town, but Mo i Rana, and even Alta or Kirkenes, might object. Certainly the only other great arctic towns—Tromsø, Bodø, Harstad and Hammerfest—all lie west of the Highway, three of them typically on islands. Narvik's present population is over 16,000 and if the suburbs are included in a Greater Narvik concept then the figure might reach 20,000. Yet, one hundred years ago Narvik did not exist (see Chapter 2).

Narvik's history as well as its future are inescapably linked with Swedish iron ore. Across the border in Sweden's Norrland is Europe's richest source of ore in the iron hills of Gällivarra and Kirunavarra. Until the discovery of the Thomas and Gilchrist process in 1878-9, the ores (60-70 per cent iron) were largely unworked because of their phosphorus content. When this problem, the removal of the phosphorus, was solved another became evident. The natural outlet of the ores was the Swedish port of Luleå on the Gulf of Bothnia. But the gulf, with its low salinity and shelter from the north Atlantic Drift, is closed by ice for half the year. An ice-free port had to be found and the nearest suitable inlet was Rombaksbotn, which reaches back from Norway's west coast to within five miles of the Norwegian-Swedish border.

In fact the site finally chosen was round the head of the Rombaks inlet where, in 1880, there were just one or two farms. The harbour was called Victoriahavn after the visit of the Crown Princess Victoria in 1887, but later reverted to the name of the farmland on which it was built, firstly in common usage and finally on its foundation as a town on 29 May 1901.

The first settlers were the railway builders who completed the line to Kiruna between 1883 and 1902. The construction of the railway was not without difficulty. Not only was there the physical handicap of the mountains, but the English company doing the work ran into labour troubles and eventually sold out its trans-shipment right. When the line was finished the Swedish route to Luleå had already been open for ten years. Political and other troubles gave rise to periods when no ore was being carried into Narvik; even the electrification of the railway in 1923 did not prevent further stoppages and consequent fluctuations in Narvik's prosperity.

However, the line is now the busiest in Norway and, with an annual capacity of 15m tons of ore and large numbers of passengers, Narvik's future might seem to be assured. Yet the town's dependency on Swedish ore must pose questions: what happens if Luleå were to be served by ice-breakers and if the trend to implant secondary industries in Swedish Lapland were to lead to a diminution in trade through Narvik?

Iron and Narvik can never be separated. Even the Battle of Narvik in 1940 was over iron ore. This wartime episode in Narvik's short history is graphically retold in the small war museum run by the Norwegian Red Cross near the town centre. The scenes, complete with life-like waxworks, are gruesome, but the museum is full of interesting detail. A small kiosk sells goods made by disabled ex-service men. A large cemetery contains the graves of hundreds of Norwegians, Britons, Poles, French and Germans who died in the battle. In Rombaksbotn lie the hulks of four German destroyers sunk in the bombing of the town.

The bombs left room for reconstruction but, although there are some fine modern buildings along Kongens Gate, the iron handling plant of the harbour, the ore stockpiles and the giant conveyors are constant reminders of Narvik's real *raison d'être*. Indeed, Narvik's preoccupation with industry and commerce leaves it without a heart.

Despite many visits, I still find Narvik a cold, unwelcoming town, but perhaps I have just been unlucky. Certainly there is much to attract the traveller. A bird's eye view of the harbour and town can be had from the mountain lift which carries one up over 2,000ft in thirteen minutes. This view is at its most magnificent in winter when the arctic sun reflects off the snow and the silvery waters of the fjord. The ski run down from Fagernesfjellet is one of the best in Europe and sports are generally well catered for. There is the largest sports hall in Norway and no less than four ski-jumps. The Norwegians take their sport—as everything else—seriously and it was in Narvik that I once watched lawn tennis being played with utter disregard of the rain storm which was rapidly turning the court into a pool.

The Arctic Highway runs through the middle of Narvik, down Kongens Gate and past Eriksen's *frihets* (freedom) monument, whose nude woman and child manage to survive the arctic winter and the summer fountains. Although the Highway bisects the town with Oskarsborg to the east and Frydenlund to the west, little of the town or harbour can be seen from the road. Even less of the enclosing mountains is visible. The famous 5,000ft *Sleeping Queen* is hidden by the parallel streets of shops and offices. The traveller who stays on the Highway has no chance to put his imagination to the test and recognise in the mountain crest the profile of Queen Victoria on her death bed.

When the Highway leaves the town it has covered over 250 miles, less than a third of its journey to Kirkenes and the Russian border.

Chapter Four

Narvik to Alta

The new road out of Narvik runs round the edge of Rombaken fjord until it reaches the narrow inner arm of Rombaksbotn. The Highway remains about 100 feet above the water as it approaches the post-war suspension bridge (see Chapter 2). Despite the fact that this is the narrowest point in the twelve-mile fjord, the bridge has a span of 2,500ft and is sufficiently high to give spectacular views up and down the fjord. On the northern side is the toll gate and the Highway eventually returns to sea level near the hydro-electric station at Trældal. The steep valley walls of the fjord have made a 570m tunnel necessary for the road west of Trældal but this excavation is less eerie than many on the Highway, for it is bright with the light of fluorescent lamps.

To cross from Rombakfjorden to Herjangsfjorden the Highway slices off the separating peninsula by rising up to 100ft through bleak bog and forest. At the end of the peninsula, now ignored, is the old ferry point of Øyjord which once served Narvik.

The head of the Herjangsfjord is reached by a straightened and widened section of road fringed by farms on a narrow lowland strip. All the way from Narvik to Elvegård and Bjerkvik at the head the twenty-five miles of Highway are surfaced with asphalt to carry the commuters daily from these suburbs into the town. But beyond Bjerkvik the surface deteriorates and, although settlement continues, the valley narrows appreciably as the mountain walls close in. At first the gradients remain low, but they gradually increase until the Highway is winding its way up the Prestjord valley to the top of the vidda. The road reaches

MAP 4

1,000ft near the border between Troms and Nordland and, although some improvements have been made, more are necessary. Hairpin bends have been straightened or in-filled with huge masses of boulders, but large white snow vehicles from the nearby military camp, as well as the normal traffic, churn up the gravel in spring.

The path to the top of the plateau is a wonderland of tree-filled gullies and mountains, with waterfalls like silver ribbons decorating their steep slopes. Down from the watershed, on the northern side of the plateau, the fall is much less steep and considerably straighter. However, snow is a greater problem here, despite the massive snow fences which have been erected. With its many lakes, the plateau is a popular place for small summer houses, wooden cabins built in the forest. At Foldvik a great waterfall passes under the Highway, and by the time the road reaches 500ft there are views of Gratangsbotn with its scattered farms perched on terraces of glacial drift cut by post-glacial streams.

In fact, the Highway never reaches Gratangsbotn. Instead, it swings eastward away from the fjord, on past the Gratangs tourist hotel and, with difficulty, cuts its way up to 1,400ft. The mountains continue to cast their shadows over the road, but the plateau is flat enough to be marshy and lake strewn, needing snow fences to keep the Highway clear in winter. A lot of work has been done in recent years on these sections of road. Some asphalt patches have been laid down and a continuous programme of road raising is planned from the maintenance station just south of Lapphaugen.

Beyond Lapphaugen, where there is a monument to Narvik's war hero, Carl Fleischer, the Highway competes with a tumbling stream to reach the bottom of the Salangs valley. It was this valley which, prior to World War I, was expected to become a second Dunderlandsdalen. But iron ore ceased to be extracted from here over fifty years ago.

Salangsdalen is forested or farmed for almost all its length. The Highway runs alongside the river, which is white with foam from its turbulent passage over a rocky bed. Great hanging valleys look down from the eastern side and hummocky moraine breaks the flatness of the floor. The Salangselv is crossed by a narrow bridge at Lund and, in places, the rocky slopes forming the valley are pinched together to give it a gorge-like passage.

To travel along the Arctic Highway for any distance is to become in danger of being satiated by the beauty of the scenery. But even the most tired eye, familiar with the grandeur of the topography, cannot be unresponsive to the loveliness of Salangsdalen. There is a unique quietness in the valley. It is another world.

At the northern end the mountains fall towards the Highway and the valley broadens. The lower mountain slopes are no longer bare but clothed in stunted birch. The floor is less used than might be expected, but some of the farms, with their greying timbers and turf roofs, are obviously old. Occasionally, a small tree grows on the roof, giving the building a slightly comic appearance. Some of the houses are on wooden stilts to avoid the marshy ground.

The Highway leaves Salangsdalen at Brandvoll, climbing away from the valley on a much improved surface. To the east is the deep valley, Kobbryggskaret, which descends from the ice-capped Lifjell. A military range is sited on the top of the divide and, as Setermoen-Bardu is reached, the military presence is much in evidence. Actually, the whole of Bardudalen has been settled for nearly two hundred years, but since the 1950s the region has become an important arctic training area for Norwegian and NATO forces.

Soldiers are everywhere in the twin towns of Bardu and Setermoen and, although the total population is little more than a couple of thousand, it is like a miniature Aldershot in its heyday. The Highway gains from the military importance of the valley, as much money and effort has been put into road improvements. The road, providing as it does a link with the larger towns of Narvik and Tromsø, is lined by hitch-hiking soldiers at weekends through the year.

The settlement is in an amphitheatre of mountains and has

important roads leading south-eastward into upper Bardudalen and Østerdalen. The Arctic Highway, however, leaves Setermoen-Bardu by taking the left bank of Barduelva and travelling north-westward in the middle valley.

The valley floor is broad enough to carry parallel roads on either side of the river. The Barduelv is sluggish in contrast to its rapid tributaries : its valley sides are plastered with glacial debris and most of the farmsteads stand like fortresses on morainic mounds. Away from the valley, the sides rise steeply, the lower slopes thinly covered with birch and pine, the upper slopes broken by hanging valleys and small cirque glaciers or snow patches. The mountain summits disappear into the clouds.

A twin-span steel lattice girder bridge crosses the Bardu river near Elverum and forms a link, via Route 87, with Rundhaug and the Målselv. New sections of road with asphalt surfaces are replacing old. Often the Highway is straightened when it is re-surfaced and the old path remains cut off from the new like an abandoned meander on a river. At Sondli a new bridge replaces the old narrow steel arch construction (*see plate, page 51*).

The village of Sondli is a farming settlement, but further on, at Andselv-Bardufoss, it is again soldiers who predominate. A large camp with a sports ground, tracked snow vehicles and married quarters proclaims this a military establishment.

It is not a very attractive settlement, perhaps because it is growing so rapidly. Nevertheless, there are some substantial new buildings, including a school, and it acts as a service centre for the whole region, which is one of the most populous through which the Highway passes. The Bardufoss civil and military airport is also an important link in the air communications of North Norway.

Reference to the early settlement of this region of Målselv and Bardudalen has already been made in Chapter 2. Much of the evidence of a *foreign* origin of the people has long since been lost, but the southern accents of the Østerdal and of Oppdal, Gauldal and Gudbrandsdal can still be detected. Even Bardu church, now nearly 150 years old, is a copy of the octagonal wooden church at Tynset.

The river Bardu flows into the Målselv near Bardufoss and the Highway then follows the latter river to a crossing by the pre-war suspension bridge Målselv Bru. From the bridge it is possible to appreciate the potential of the valley, which encouraged the eighteenth-century settlers. Not the least attractive feature today is the salmon which leap up the ladders round Bardufossen to reach the mountain lake of Lille Rosta. From the bridge, too, the great 4,000ft Mauken range can be seen to shelter this valley from the north west.

The Mauken range forms one side of the mountainous col which the Arctic Highway uses to reach the Balsfjord some twenty-five miles east of Olsborg. On the other side, Breitinden and the Slettefjell rise about 250ft higher. With a sharp Z-bend the Highway leaves the Måls river valley and makes for the intermontane col and plateau. For a while the road has an easy passage along the valley of the Takelv, which has the same type of old farm buildings as those in Salangsdalen. However, towards the head of the valley, more open forest and peat bog replace the farms and the area around Bjørkli and Talkelvdal is another military training ground. The older road, away from the river and previously used in time of spring flood, is now abandoned. The open plateau at the watershed is only 600ft above sea level and consequently has been settled. Some of the most attractive sites for farms are around a large and placid lake, Takvatnet.

At the northernmost corner of the lake is Heia and the junction with Route 857. This road gives a link with the upper Målselv and Tamokdalen. Because of the relatively heavily settled character of this part of North Norway, there is a quite unusually close network of roads which gives the traveller alternatives to the Arctic Highway.

By the side of the Highway at Heia is a granite monument to the engineer Paul Holst and nearby, in summer, a group of Lapps pitch tents among the rocky outcrops and bogland.

Yet another large lake lies on a lower platform of this plateau. At an elevation of about 270ft, this is Sagelvvatnet and it feeds the river whose valley the Highway now uses to descend to Balsfjorden. Much of this part of the country is quite open, with a pleasant, well-ordered human landscape blending with forest and mountain streams.

The descent into the important Balsfjord is quite gentle. Just before reaching the village of Storsteinnes, the giant Markenes-dalen can be seen to hang above the level of the fjord. The village is well sited. It lies sheltered within a little bay called Sørkjosen which indents the side of the fjord. It has the protection of a small headland to the north and eastward lies the Sagelvdalen, through which the Highway has reached the settlement. Communications round the fjord shore are better than adequate on the talus platform, which also supports a string of farms.

Today, Storsteinnes is a small fishing port and trading centre for this upper part of the fjord. It is a neat little village with a sizable dairy. From the village, the Highway turns sharply eastward to run along the edge of Balsfjorden. All around is the mountain scenery for which Tromsødistriktet is justly famous: sharp peaks, sheltered cirques and U-shaped valleys of wonderful concavity. In summer, snow patches or small cirque glaciers occupy the hollows; in winter, the angular peaks thrust their bare summits through the mantle of snow.

At Markenes the Highway leaves the broader Balsfjord to penetrate the inner arm of the mis-named Nordkjosen. The valley sides here are especially steep on the southern side of the inlet, with peaks rising over 4,000ft and the greatest, Istind, nearly 5,000ft high. Clouds form easily and, with temperature inversions, frequently hang like garlands along the mountain slopes. At other times the peaks sprout banner clouds clear against a blue sky. Across the water the important Tromsø road can be seen following a parallel course on the opposite shore.

At the head of the inlet, Nordkjosbotn, is the village of Vollan. This is an important communications node, for it is here that the Arctic Highway meets Route E78, the main road to the Fylke capital of Tromsø, and also where traffic from the Finnish panhandle uses the Highway as its international link. The village is usually a bustle of activity in the short tourist season and benefits from its proximity to Tromsø. The fjord here is in-filled with river and glacial deposits, forming a flat bay-head which is gradually extending into Nordkjosen. As reclamation takes place, so the village migrates westward.

Behind Vollan is one of those rare low-lying passes between the mountains which the Arctic Highway is quick to use as its path. This particular defile is only twelve miles long, but it conveniently links Balsfjorden with the great Lyngenfjord. The Highway has to rise only just over 300ft, at Øvergård, to cross the watershed between Nordkjoselva and Balsfjordeidet.

These two valleys are relatively broad, their flatness only interrupted by eroded drift. A few farms, brightly painted wooden buildings on concrete bases, are scattered along the road where the valley widens. To the south, mountain cusps continue to break the sky-line like bared teeth, but the smooth-sided valleys between them are of incredible beauty. Few lead anywhere except to a blind head, but they are nonetheless inviting. The naked sides of the peaks and of Kilafjellet to the north are gullied with torrential streams on the higher slopes, but the lower parts are clothed in birch. The north-south contrast shows the expected valley glaciation of the north-facing aspect. Indeed, this side keeps its snow all summer.

Just before Øvergård there is probably what is the most bizarre feature of the whole of the Arctic Highway's length. On the north side of the road, half hidden by trees, is *Piggsteinen* (the pointed rock). Standing about 12ft high, this erratic boulder has become a sort of travellers' memorial. As a change from the usual graffiti which, along with litter, seem to be the average tourist's contribution to the countryside, this rock is decorated with names, initials, dates and slogans painted on the smooth face. Quite where the coloured paints come from I have never discovered. I half suspect that a sign writer from the Tourist Board in Oslo travels north occasionally to attend to the work and give the impression of international tourism along the Highway! Anyway, there it is, a record that *B-G J* travelled this way on his journey from Paris to Cap Nord and that *PCS* from Kiruna stopped here en route for Tromsø.

At Øvergård, Route 87, which left the Highway in Bardudalen to serve the Målselv valley, catches up again by way of the Tamokdal. This is a most interesting alternative to the Highway and, in fact, is something of a short cut (see Chapter 6).

At the head of the Lyngenfjord, an arm called Storfjorden, the Stordal river meanders through terraces. Just before the head is reached, the international road into Finland, Route E78, parts company with the Highway to take the eastern shore. Soon this road will be part of the Arctic Highway when the Skibotn-Laukvoll link is complete (see Chapters 6 and 8).

From the Highway the great U-shaped valleys of Signaldalen and Kittdalen can be seen framing conical mountains. To the south the twin pyramidal peaks of Otertind rise over 4,250ft. Otertind is usually described as Norway's Matterhorn and the Signal valley is an attraction to summer tourists based on Tromsø. The region has a relatively high snowfall in winter and it is this which helps to sustain the cirque glaciers of the Lyngen Alps.

The whole length of Lyngenfjorden and its inner arms are dotted with tiny fishing-farming communities, hamlets of a dozen or less houses. There is little space for building or for fields, so sites have been chosen where delta flats reach into the fjord. These great fans of alluvium have been terraced by the rivers which laid them down and the farms stand on these ledges like dolls' houses in a toy shop display. On one side of the road, in summer, hay is set out to dry on wire fences and on the other side, on the beach, fishing nets and racks of fish are similarly arranged to catch the sun in the long daylight hours. The rivers are reduced to a trickle in winter, but in summer they discharge their milky melt water into the dark depths of the fjord.

From Rasteby, on the Highway, the fjord widens; on the eastern shore is the old market village of Skibotn at the mouth of the open Skibotndalen and below the commanding slopes of Falsnesfjellet. During the 1940-5 war the Germans sited an artillery emplacement on this mountain using Russian prisoners to build a service road up the steep sides. Precipitous slopes are characteristic of Lyngenfjorden. Rastebyfjellet and Pollfjellet tower like great walls above the Highway, obscuring their own glacier-capped summits by the very steepness and convexity of their faces. In spring, avalanches are common along the Highway between Rasteby and Kvalvik. For a month or two each year the Highway may have to be closed and a ferry links Sandvika with Lyngseidet. Snow and rock are liable to thunder down on to the road without warning, but the local people seem to know the signs and loss of life is rare. However, the story is told of two German soldiers who, during the war, stayed to take photographs of the falling snow and were killed by their folly. The occupying troops brought in a firm from Germany to build an avalanche shed over the Highway, but this met with no success. In summer the only signs of danger are the large scree fans and the thin streamers of waterfalls which spill off the mountains.

The village of Furflaten is much like its neighbours, Elvevollen and Rasteby, but the terraces into Lyngsdalen are somewhat higher and the village jetty rather larger. Further north along the Lyngenfjord, Pollen and Kvalvik occupy a different sort of site. Here, sediments have filled the narrow strait between the shore-line and a rocky outcrop which would otherwise be an island in the fjord. Pollen and Kvalvik lie at each end of this alluvial *bridge* which carries the Highway. On the *island* are the hamlets of Flatvollen and Sandvika which are to be given a link with the Arctic Highway.

The coastal terrace widens north of Kvalvik and, looking back south-westward from the Highway, the great height of the mountains can be appreciated. This southern rib of the Lyngshalvøy peninsula contains the highest peak in North Norway: Jiekke-varre, which is over 6,000ft. The peninsula is almost sliced into two by the Kjosen arm of Ullsfjorden and it is at this break that the Arctic Highway reaches Lyngseidet, its ferry point for the crossing of Lyngenfjorden.

On the approach to Lyngseidet there is a fine wooden school and the whole village has an air of prosperity, order and purpose.

In fact it has an important history (see also Chapter 2) and was established as a trading centre in 1789. Its church is even older. The original building dates from 1731, but it was extended fifty years later and, in contrast to so many North Norway churches, given a cruciform shape. Today, the village holds not only the important position of ferry point, but is also at the junction of the Arctic Highway and Route 91 which leads to Tromsø (see Chapter 6).

Lyngseidet is very much a communication centre and meeting place. The village is often crowded with people and never more so than on a summer Sunday. Lapps, who camp nearby, come into the cafés and mingle with churchgoers from the surrounding hamlets, tourists from Tromsø and the Arctic Highway travellers waiting for the ferry. It is sad to think that some of the bustle and cosmopolitan atmosphere of the village may be lost when the Highway forsakes the ferry on the completion of the Skibotn-Laukvoll link road.

Although the motor vessels *Haalogaland* and *Jœggevarre* are the largest of the Arctic Highway's ferries, there are frequently long queues of vehicles in summer. The ferry takes the Highway across the Lyngenfjord to Olderdalen on the eastern shore in thirty-five minutes, but this is a very significant half-hour journey.

From Olderdalen the Highway has no further need of ferries; its path is eastward rather than to the north and the landscape becomes more obviously arctic. On leaving the Olderdalen quay one senses that there is no turning back. One is committed to complete the length of the Highway: this is the point of no return.

Olderdalen is much smaller than Lyngseidet—just a café, a guest house, one or two shops and a handful of houses. From the road, the great Lyngen Alps provide magnificent scenery on the other side of the fjord. In winter their 5,000ft peaks peer through the dull arctic twilight, but in summer they can be seen in all their glory, with snowy caps and glistening ice-blue cirque glaciers. The mountains continue to attract foreign mountaineers, especially from England, but many faces have yet to be scaled.

Along Lyngenfjorden, on this eastern side, on narrow terraces is a string of small farms. Little land is suitable even for hay, but, by combining fishing with farming and using the coastal strip right down to the water's edge, the people manage to make a living. Indeed, the settlement is almost continuous along the shores of the Lyngenfjord and Rotsundet. During winter, fishing in the Arctic Ocean or the Norwegian Sea is the only occupation, but, in summer, the goats and cows which have been stall-fed while snow was on the ground move to the higher pastures and hay is harvested in the small fields by the farms. Huge racks are erected on the beach to dry the fish, for there are, as yet, no readily accessible freezing or canning stations. The local market is small, but it is not uncommon to see fishermen going to the villages along the Highway on bicycles festooned with fish just landed.

Where rivers empty into the fjord, their deposits and those of the glaciers which preceded them give wider terraces on which small villages have grown up. The mouths of the valleys are frequently blocked with moraine which is cut into by the rivers, giving a small gorge through the drift. In these instances the Highway has to abandon the higher terrace and slip down to the shore to cross the river by a narrow bridge. Three villages—Nordmannvik, Engnes and Djupvik—face west into Lyngen-fjorden, and Rotsund looks across Rotsundet towards the island of Ul. Uløya presents a steep forbidding face to the mainland and the dangerous waters of the sound run fast and deep. The Rotsundal is farmed along its length, enjoying the shelter of the island from the northerly, arctic winds.

It is important to remember that the coast along which the Highway is built is, from Olderdalen, largely aligned east-west and therefore facing north towards the Pole. On-shore winds are bitterly cold in all seasons and the warmth of the North Atlantic Drift is less effective in moderating the temperatures. Moreover, as explained in Chapter 1, the fjords of the extreme north are wider and less steep-sided, thus allowing the easier passage of air from sea to land.

From the Arctic Highway at Landsletta, there are views of the other islands which, with Uløya, guard the north coast. Most rise steeply to heights over 1,000ft and, except on especially warm summer days, are shrouded in cloud. Beyond Landsletta, the road leaves the coast to cross the narrow neck of a peninsula. Here the Highway climbs to over 700ft through thick birch wood. Near the top of the rise is a road maintenance station and there are Lapp camps in summer. These are *real* Lapp camps, unlike that set up some miles back on the Highway on the Djupvik ridge where a group of Tromsø Lapps drive out by car each day in season to sell skins and trinkets to the few tourists who pass by.

From the peninsula, the descent to Reisafjorden is steep and winding. By the fjord shore are two quite important villages: Sørkjosen and Nordreisa. The former has an hotel which is used for one of the overnight stops of the North Norway Bus and the latter is an expanding fishing village. Much of the settlement is on the reclaimed mud flats of a delta which is building out into the fjord from the waters of Reisadalen. Already Nordreisa incorporates the previously separate hamlets of Storslett and Solvoll, and Sørkjosen will soon be absorbed. In the future (see Chapter 8) the likelihood of an interior link road off the Highway towards Finland suggests that the importance of this settlement will increase even more.

Today, Route 865 leads down the wooded Reisdal, a valley largely inhabited by Lapps or Kvänsk-speaking Norwegians of Finnish origin. Beyond Bilto it is possible to continue the journey up the valley, but not by road. Instead, a boat takes travellers up to the huts at Nedrefoss. This is a rewarding excursion, for one passes the great 900-foot Mollesfoss waterfall and salmon are plentiful in the Reisaelv.

However, the Highway ignores such diversions and presses on along the coast round the fjord and through the hamlet of Flatvoll. Another peninsular neck is crossed to reach the next inlet, but a rise of 100ft is necessary. Along Straumenfjorden the Highway occupies a ledge only 10ft above the water.

Running between a mountain wall and the green waters of the fjord, the road rises and falls alongside Straumenfjorden and

Oksfjorden like a rather half-hearted switch-back. To the north-west are the islands of the Skjervøy group and ahead is the Troll-dalstind. The grey-pink 3,600-foot Nuovas peak dominates the view.

Here and there small fishing communities exploit the inshore waters, rich in haddock and cod. As elsewhere, they supplement their income by engaging in small scale forestry and dairy farming, but, without the advantage of a processing plant and with a diminishing market for dried fish, one wonders how long it will be before the harshness of their life drives them south or into the towns like Tromsø or Alta.

The natural head of Oksfjorden is avoided by the Highway. Instead, the road is built across a moraine which effectively cuts the fjord in two and then bridges the outlet of Oksfjordvatnet which, in fact, is the inner arm of the fjord. This lake has a unique beauty, enclosed on three sides by high mountains. As the road climbs steeply away from Sandbukt by the lakeside, the hamlet of Mettevoll looks like a toy village set at the water's edge.

Now begins one of the Highway's more difficult sections. Up to the middle 1960s the road here was closed through much of winter, but now, except for perhaps a week or so, the Highway functions regardless of the weather. The problem is largely one of altitude and exposure, for, in avoiding the long coastal route round the Trolldalstind range, the road has to climb on to the Kvænangsfjell some 1,300ft above sea-level. This would be difficult enough for any road, but, at almost 70°N and with the eastern limits of the plateau open to the north and the Arctic Ocean, the audacity of the engineers can only be admired.

Long closures have been eliminated by the usual expedients of ploughs, snow fences and surface raising. In winter one is likely to be confronted by a barrier across the road carrying the instruction 'Stop! Wait for the snow plough'. But the wait is not long and, between walls of snow, the winding ascent to the top can be made. At 1,000ft the shelter of the Trolldalstind is lost and thin birch gives way to open peat bog. Massive snow fences, some with wire mesh, have been set on the plateau and the road reconstruction programme has included the building of crash barriers, as well as widening and straightening the road and raising it between four and eight feet.

A road maintenance station is at the summit; so, too, is a collection of turf gammer (*see plate, page 157*) belonging to the Lapps who graze their reindeer here each year (see Chapter 7). Even in summer, Kvænangsfjellet is frequently enveloped in swirling cloud, and snow may fall in any month.

The Kvænangsfjell is more of a ridge than a plateau and the Highway drops quickly off the top, down towards the long fjord which bears the same name. The fjord is wide and its sides less steep than those of southern fjords, but it also gives less shelter. Cold arctic air blows across the Highway as it descends towards the sea and birch trees do not appear again until 700ft, and then only where there is shelter. At 500ft there is another snow plough barrier and the Highway begins to level out above the fjord. The road reaches the fjord edge at 300ft.

Beside most of the western shore of the Kvænangsfjord and its arms, the Highway uses a narrow ledge cut out of the mountain slope. On the seaward side of the road the drop is always precipitous, but on the other side a more gentle birch-covered slope in places separates the Highway from the steep bare rock faces of Gaggavarre. At some points the road carves through thick scree and it is difficult to distinguish between boulders dynamited in the construction of the road and the natural talus.

The road has to cross many streams which tumble into the fjord; consequently, it follows an undulating path, sometimes down almost to sea level, and at others 350ft above the fjord. Mostly, however, it is at 150ft and affords clear views of the eastern shore. The narrow width and water-bound gravel surface of the Highway are additional difficulties. In summer, the giant scrapers (*see plate, page 33*), which swallows the gravel at the front and redistribute it through combs at the rear, are hampered in their task by the necessity to retract the combs to allow vehicles to pass.

At Karvik, where Arriselva enters the inlet, the mountains

retreat back from the fjord and the lower valley is filled with cotton grass in the warm season. It is here that the old ferry used to cross and where a new bridge is to be built (see Chapters 2 and 8 respectively).

Just before the Highway leaves the fjord to drive across towards a hanging valley and to Sørfjorden arm, an impressive sixty-foot drop waterfall, Navitfoss, crashes under the road. Away from the fjord there is a scattering of Lapps' huts, some quite recent. These are substantial buildings, the equal of those belonging to Norwegians in the villages. The Lapps have settled here from the Kautokeino district. Their nomadic wanderings are over.

The Arctic Highway falls to the very edge of Sørfjorden, runs along it scarcely ten feet above the water level and crosses a narrow peninsula into Kvænangsbotn at the fjord head. Kvænangsbotn has the advantage of a new hydro-electric station and a small timber industry, but the farming is poor and fishing, as ever, is the mainstay of the small population. The café and shop will doubtless suffer when the Highway bridges the middle fjord and quits Kvænangsbotn.

The fjord head is surrounded by the denuded slopes of Čorrovarre, Rossavarre and Orddavarre. Their rock surfaces are free of any soil and clearly show glacial striations. Rocky islands litter the fjord which is subject to considerable sedimentation.

Beyond Nordbotn the Highway rises away from the shoreline, occasionally losing sight of the fjord altogether, but never very far from it. The road reaches 350ft before dropping steeply again to the fishing hamlet of Kjækan which boasts a strong jetty. This eastern side of the fjord is very desolate. Craggy, truncated rock promontaries give way to bare low plateaux which lead up to the higher slopes of Orddavarre. There are few trees and fewer farms. Every farm has its own boathouse, as if to demonstrate its dependancy upon both land and sea. These communities are some of the most isolated in the whole of Troms.

The fishing hamlet of Sekkemo will benefit from the proposed fjord bridge, but at present it simply marks the Highway's

Page 121 (*above*) Tromsø, a view to the east across Tromsesund. The bridge, with its 3,400ft span, was opened in 1960. Tromsdalen church on the mainland is said to symbolise the polar night and *Aurora Borealis*. (*below*) The Lapp village Kautokeino in late spring when most of the winter's snow has melted. The church in the centre background was built in 1958 close to the site of the original 1703 chapel.

Page 122 Cars queue at Svensby for the Ullsfjord ferry on Route 91 to Tromsø. The canoe on the car in the foreground carried the author into otherwise inaccessible parts of Lapland.

entrance into the drift-filled Badderfjord—another inlet off the main fjord. This embayment is well farmed on the richer drift all the way through to Undereidet where the Highway cuts off a further headland by way of Baddereidet.

This pass provides varied scenery as the road rises through it. Most is utter wasteland—firstly rocky slopes supporting gorse or thin birch, then treeless bogs as the Highway reaches its maximum altitude of 850ft and snow fences are necessary. On the descent on the northern end of Baddereidet, the path is abrupt and partly wooded, but bog persists to 300ft. Nearer to Kåsen the gradient lessens and the side of the Ridevarre mountain appears to have been hacked away by some Brobdingnagian chopper, its bare rock face crossed by sharp horizontal ridges.

Burfjorden is the next inlet whose shore provides the Highway with a path. This is a sheltered fjord and Burfjord village stands at its head, providing a market place for the surrounding country. Although it has taken over this rôle from Alteidet at the mouth, Burfjord, too, has had its setbacks, for the copper mining in Burfjordalen has long since ceased.

Although fishing is now the most important occupation, some cattle and sheep are kept by the families living along the fjord. These animals add to the difficulties of drivers on the Highway as they wander along the road, bells ringing, apparently unaware of danger. There is a large road maintenance station in Burfjord, but Alteidet boasts a bank and a large store, and its old buildings are altogether more attractive. This latter village is reached by the Highway following a path which skirts the edge of the fjord almost at water level. Its importance as a trading post in the past was partly due to the use of the Alteidet valley by migrating Lapps each summer. Today the valley is farmed, despite its rather marshy character, and the shallow slopes are tree clad. Near the edge of the pass a plaque mounted on a monolith proclaims that the Highway is leaving Troms to enter the most northerly fylke: Finnmark.

Finnmark is the traditional meeting place of northern peoples: Lapps, Finns, Russians and Norwegians. In the early nineteenth

H

century, Norwegian sovereignty was finally established although the Norwegians were latecomers to the region, their first settlement probably dating from the twelfth century.

The Highway takes a sharp turn to the south at the eastern end of Alteidet. To the north is the pretty Russelv valley and ahead is the Langfjord. This inlet is remarkable for its length, straightness and narrowness. It is possible to see the whole of its twenty miles from vantage points at the head. Sedimentation is taking place rapidly at Langfjordbotn and shallow flats are uncovered at low tide.

The only settlement of consequence is near the head at Bognelv; there are the workers' houses at the nearby road maintenance station and a little market gardening is practised, taking advantage of shelter from the north in this east-west aligned fjord. All the same, farming is generally poor and the land can give only partial support to the few families who live here.

The quite extraordinary perseverance of the people in this harsh environment can be seen on the opposite side of the fjord. Parallel to the Highway, on the southern shore, the northern coastal strip, with its slightly more favourable aspect, carries a string of tiny farms. For almost three quarters of the length of the fjord, hay is grown on seemingly impossible slopes and later put out to dry in conditions of quite unreasonable humidity and precipitation. But beyond Riverbukt the mountain side becomes a vertical cliff and the challenge can no longer be accepted.

The Highway's undulating path south of the fjord is unhampered along a desolate course. As so often before, the road rises and falls with the fretted shore, sometimes travelling up to forty feet above the water's edge and then down at sea level again. The slopes of Lassefjellet, of the Langfjell and of Høyfjellsnosa threaten the Highway on the landward side. The lack of sun leaves the slopes bare even of scattered birch above 300ft and enormous screes rest as monuments to freeze-thaw weathering.

At Ulsvåg a stream off Høyfjellsnosa has cut a deep series of terraces into the morainic waste, blocking its entrance into the fjord and, near Eidsnes, a striking gorge from the same mountain gushes water into the inlet by way of a fall. The Høyfjellsnos, as its name suggests, is very steep and in winter avalanches are a real problem. Snow fences are often inadequate and need frequent replacement.

Before the Highway rounds the headland separating the Langfjord from Altafjorden, one can just see into the forbidding Ytre Kåven on the opposite shore. This mountain-girt cove breaks the otherwise regular and vertical cliff line. Once into the broad Altafjord, the contrast with Langfjorden is manifest.

The Highway climbs 200ft to enter this fjord and a superb viewpoint is provided by Isnestoften promontory overlooking Langenesholmen. The Germans were quick to appreciate the advantage of this position during the war and their artillery emplacements can still be seen in the complex network of platforms, enclosing walls and bunkers cut out of the rocky mound. They were not the first to settle at Isnestoften (or Toften, as it is usually called by the local people). There is evidence that this was a neolithic dwelling place and further excavations suggest that Lapps settled here about 1,000 years ago.

Perhaps the most picturesque view from this headland is that towards the head of the fjord in mid-winter, when, in the half light which is the arctic's shadowy illumination, the twinkling lights of Alta announce that this wasteland is not without a haven. But Alta is still over forty miles away by the Highway.

Most of the western shore of Altafjorden is uninhabited, except for a solitary farm or group of farms sandwiched between fjord and mountains. The road here is narrow and rather dangerous, with its usual undulations and some especially sharp bends. From the highest points on the road, up to 200ft above sea-level, the eastern coastline looks more uninviting, even sinister, as blue-black cliffs plunge to the apparently bottomless waters of the fjord.

A small bay provides shelter for a rather larger community at Talvik. This little village, at the mouth of the river Stor, was previously a trading post for the whole of the fjord, but its recent incorporation into Greater Alta seems to have set the seal on its declining influence. The bay itself is surrounded by relatively low hummocky hills of doubtful origin and the village has a fine church as well as a bank and the customary store. Alta is hidden from view at Talvik but reappears as the Highway goes around the promontory at Flatstrand and heads towards the Kåfjord arm of the fjord. The road first crosses a low col between Storvik and Flintnes, which is used as a base in summer by reindeer Lapps from Kautokeino; then it falls towards the fjord's edge.

The Altafjord has given shelter to ships since Viking days, but the most famous vessel to use the Kåfjord was the German battleship *Tirpitz*. This great ship, together with others of the German Atlantic Fleet, hid here in the last war. Because the inlet is filling in rapidly with glacial and river deposits which have almost cut the fjord into two, the *Tirpitz* was forced to use the outer part of the fjord arm. The high mountains gave protection from air attack, but the deep water at the fjord entrance allowed British one-man submarines to enter, guided by Norwegian-manned radio transmitters on shore, and cripple the vessel. Its hide-out discovered, the *Tirpitz* limped south to Tromsø with over 10,000 men on board and was finally sunk in a bombing raid in November 1944.

Kåfjord's earlier fame had come from the copper workings in the mountains to the west of the village. The hill slopes are peppered with old mines, but there has been little activity since the mining company left the fjord in 1878 after a half century of exploitation. The company (called the *Altens Kobberverk*) was founded by two Englishmen, Crowe and Woodfall, in 1826 and later taken over by a Mr Robertson, whose descendants still live in Finnmark. In fact a number of Englishmen were amongst the 500 workers who mined the copper; one, a Mr Thomas, had the unique distinction of being elected to the Norwegian parliament, the Storting, much to the jubilation of the Kåfjord community. In the celebrations that followed the election, it is recorded, much champagne was drunk. The British also erected the church, the second oldest in Finnmark, and were generally more welcome here than in the other regions of North Norway to which they were led by mining interests.

Today there are only a couple of hundred people in the area and the massive Kåfjord Kobberverk and old jetty lie derelict, an untidy scar. When the copper began to be worked out, many families emigrated to the USA.

A crumbling path leads out of the village through the mines and up to the Haldde range. Just over a mile from the end of the path, perched on the 2,969-foot Sukkertoppen peak, is the old *Nordlysstation*. The world's first aurora borealis or northern lights observatory functioned here at the turn of the century until its transfer to the new auroral observatory at Tromsø.

Across Kåfjorden the sheer walls of Sakkovarre let loose their scree in a cataract of rock, but the Highway avoids this eastern shore. Instead, it crosses the fjord by a causeway through the sedimented head and makes towards the broad Mattisdal. Both the Mattiselv and Botnelva have incised themselves into the sediments, leaving terraces of drift and alluvium to mark successive levels of erosion.

The Arctic Highway does not go up Mattisdalen, but spans the river with a simple bridge and swings northward behind the less precipitous eastern slopes of Sakkovarre. There is a very minor road up the valley and, for the fishing alone, the Mattiselv is worth a visit.

Alongside the Highway is wooded, but the steeper slopes above are bare. Although the road surface is still rather poor here, the gradients are low, for this section is built across drift and deltaic deposits which block what ought to be another long arm of the Altafjord. The two lakes, Storvatnet and Kvænvikvatnet, and their marshy surrounds were once a continuation of the small Kvænvik inlet. The Highway does not quite reach the bay, but cuts across the bogland to rise 200ft on to the side of Skaddefjellet and eventually rejoin the main fjord at Simånes. This village lies below a huge moraine which shuts off the valley of a river feeding into the fjord.

From here the Highway makes straight along the border of the fjord into Alta. Just before the town is reached there are splendid views north down the fjord and of the rock flats and bays of

Alta. What cannot be appreciated is the size and scattered nature of the settlement, although its importance might be judged by the new asphalt surface of the Highway and by the street lighting.

Chapter Five

Alta to Kirkenes

Like Mo i Rana, and probably for the same reason, Alta is remarkably little known. Although on the Arctic Highway, the little
town—it cannot still be called a village—is not a port of call for
the *Hurtigrute*. Furthermore, it is less a Norwegian settlement
than one of the Lapps' or, more especially, of the Kvæns'. Alta
is quite unlike any of the other larger arctic settlements, for it
owes its importance and its origin uniquely to the land rather
than to the sea. It is a Lapland town in a way that cannot be
compared with Hammerfest or Kirkenes, Vadsø or Vardø. Alta
has a greater affinity with the interior Finnmark Lapp villages
of Karasjok and Kautokeino than with other fjord settlements.
Moreover, of all the larger settlements on the Arctic Highway,
Alta is most dependent upon the road and for this reason
demands special attention.

Some of the explanation for this singularity lies in the climate.
At first sight there seems little in Alta's temperature statistics to
contrast it with other coastal stations in North Norway but, on
closer inspection, two important facts emerge. Firstly, the maximum summer temperatures generally exceed those recorded
elsewhere in Arctic Norway. Secondly, but more importantly, the
warm season also lasts longer, giving the benefits of a longer growing season as well as higher accumulated temperatures through
exceptionally high daily maxima (see also Chapter 1). Admittedly,
the differences are not of a high order but, in a region where even
half a degree or an extra week of warmth is significant, the importance should not be underestimated. As in most of NW

MAP 5

Europe, recent decades have shown a significant lowering of temperature compared with the first thirty-five years of this century, but the contrasts between Alta and other arctic settlements remain.

The precipitation at Alta is lower than in much of Arctic Norway and, although figures are not as low as those recorded in interior Finnmark, the annual total is such that the winter snow presents a relatively smaller problem while, in summer, rainfall is adequate.

Two other important contributory factors help to make Alta's situation unique. Of prime importance to a community that looks to farming rather than fishing for its main occupation, the land around the head of Altafjorden is well suited to cultivation. Much of the Alten region has been fashioned by the great Altaelv and by the smaller Tverrelv and Transforelva. Apart from sediments laid down by these rivers, the terraces they have cut in the glacial drift which chokes their mouths favour farming on a scale unparalleled in Finnmark and rarely surpassed in Nordland or Troms.

Lastly, the Alta river gives access to the Finnmark vidda from which the early settlers came and which, even today, is the town's natural hinterland. Not until the Tanaelv is reached in East Finnmark is there such a natural route into the innermost parts of Lapland.

The earliest people to occupy the site did so some eight thousand years ago. In 1925 the archaeologist Anders Nummerdal established the presence of stone-age dwellings on the Komsafjell, a promontory which juts out into the fjord in the middle of the town. These Komsa people, with their paleolithic culture, are thought to have come from the region about the Gulf of Finland as the Quarternary ice sheets ablated. They may have been the progenitors of the Lapps (though many dispute this) and amongst the finds have been what are thought to be primitive skis.

Little is known of settlement in Alta between these earliest times and the sixteenth century when some fishing settlements arose around the fjord. The concentration at the head and the establishment of nucleated villages dates from the beginning of the eighteenth century. In 1703 the Alta Kommune was set up to encompass much of the shore of the fjord including Talvik. Almost simultaneously, large numbers of Kvæns (also: Quains or Quäns), driven out of Finland by taxation and political unrest, descended into the region from the vidda. These people, with their distinctive language and culture, have left their mark on the town. It was they also who, with their non-littoral origins, brought agriculture to the valleys and introduced grain cultivation. Their descendants can be distinguished by their accents, if not by the Kvænsk language, and by their Finnish names.

For decades the region was Finnish and Lappish rather than Norwegian. The houses were large Finnish log cabins with a single room in which the furnace heated the *sweating baths* (sauna) above. The Norwegian language was scarcely heard and even in 1801, of the 1,800 people in the extensive Alta parish, less than 500 were Norwegian. Alta owes a lot to the cosmopolitan origins of her inhabitants. Mining in the area in the 1830s, by the Dutch and then the British, introduced large numbers of workers from the Dovrefjell in Sør Trøndelag as well as a few foreign administrators and technicians from the mining companies. Even today many of the inhabitants are from the south and the *sør trøndere* have become the farming élite.

As Alta's importance grew, that of Talvik declined, for, in keeping with its rural origins, Alta became a great market centre for West Finnmark and from 1791 was officially recognised as such. The official market moved from Elvebakken to Bossekop in 1844, but the latter was already the most important Lapp trading centre in all Fennoscandia. The great December and March Lapp fairs ceased here only after the last war.

None of the old buildings, except the church, is left. Gone is the old Altagård, the seat of the district magistrate or amtmann (*amt* was the old name for fylke); gone, too, is the Royal *mansion* which gave its name to the Kongshaven, on which shore it stood. No longer is there *Vina*, the villa where, from 1862, generation of the English Dukes of Roxburghe spent their summers in pursuit of the salmon of Altaelva. In 1944, the retreating Germans destroyed the town by fire.

The war brought hard times to this part of Finnmark. The Alta battalion, then recruited entirely from this area and including some of the most prominent citizens, brought credit to the town in the battle of Narvik, taking no less than 200 prisoners. Later the people of Alta had to act as hosts to the occupying power, but the cruellest blow of all came towards the end when the population—less some forty who took to the forests—were forcibly evacuated by sea while their town was set aflame.

Now there are new buildings, almost exclusively wooden but brightly painted as in all North Norway. The people have returned and rebuilt their town, overcoming this disaster just as they overcame the ravages of a mysterious disease which wiped out a large proportion of Finnmark's population in 1806. The population today numbers some 11,000 in the Kommune, of whom 6,000 live in the town. By 1980 an increase of 4,000 is expected as the new *city* is built. The plans for this are briefly discussed in Chapter 8.

Even in its earliest days, Alta impressed the visitor. That most observant of Lapland travellers, von Buch, wrote in July 1807: *How beautifully rural Elvebakken appears at the mouth of Altens Elv! It looks like a Danish village. The houses, to the number of about twenty, lie up the banks of the great stream in the midst of green fields and meadows, surrounded with high Scotch firs in every direction. Alten is not only the most agreeable, the most populous and the most fertile district in Finnmark but also the only place in which agriculture is carried on—the most northerly agriculture in the world.* Well over a century and a half later, Alta remains the most attractive major settlement throughout the length of the Arctic Highway.

Perhaps von Buch would not recognise the town today, although the trees *so beautiful and so diversified* remain and the saw mills still run. But the three small villages of Bossekop (Whale Bay), Bukta (Bight), and Elvebakken (River Banks) have coalesced into a single settlement: Alta. Much of the rural character, however, remains, for there are not the same physical restrictions to the site as are found so often in fjord settlements. Only from the air can the dispersed nature of the town be appreciated, houses set among the trees and only the Highway to give them a focal axis.

Because the three villages, Bossekop, Bukta and Elvebakken, once enjoyed a separate existence, there is no real *sentrum* or core to the town. If it has any shape at all it is linear, strung out along nearly four miles of the Arctic Highway. Near the western entrance to the town is the only hotel, smaller and less imposing than one might have expected for a town of this size. The North Norway Bus stops here and when it arrives, its passengers crowd the cafeteria. Lapps, Norwegians, Finns and summer tourists queue for their *middag*.

Down the hill from the hotel the Highway runs through Bossekop. The crossroads, where Route 93 leads south off the Highway towards Kautokeino and a path leads down to the bay, is the nearest that Alta comes to having a town centre. Here are the post office and telephone exchange, a bank and a garage, but most especially, here is the main store. This store is a meeting place for the Lapps in summer and its entrance often appears more like a club than a shop (see also Chapter 7). From Bossekop the Highway climbs again behind the Komsafjell promontory, past the wooden cinema, where the audience waits more or less patiently while the reel is changed on the single projector, past a café which serves the best *lapskaus* in Arctic Norway and down a steep hill towards Bukta. In the eastern bay is a seaplane base—an anchorage off the rocky shore—and at the bottom of the hill is the barracks of the Alta Battalion.

The best salmon river in Europe (some say, in the world), Altaelva discharges into the fjord at Elvebakken and its sediment, forming a large delta, has been reclaimed for the site of Alta's airport. The river itself is crossed by a graceful suspension bridge and the Highway leaves Alta soon afterwards by the Tørfossen Bru which spans the Tverrelv.

There is an indescribable charm about Alta that makes it difficult to leave. Perhaps it is the very rural setting, farms within the town and houses set down in pine and fir woods. Little industry is visible; the slate quarries producing the famous *Skifebrudd* tiles are outside the town and the cleanliness and order of the whole settlement are striking.

With Alta behind it, the Highway continues round the fjordhead, first away from the shoreline, hidden by trees; then, after passing the mouth of Transfordalen, along the edge again with good views across the fjord and back towards the town. The road surface has been improved on this section and realignment and widening have taken place. The inlet which is followed by the road is Rafsbotn, an easterly arm of Altafjorden which, with Kåfjorden, gives the whole fjord the shape of an inverted T.

At the hamlet of Rafsbotn the Highway turns inland away from the headland formed by Altenesfjellet. A narrow road leads off to the west to travel round the fjord for another five miles, past broad weed-covered beaches and an occasional farm, to Russeluft. The Highway, however, climbs up through a gorge away from the fjord. Birch trees manage to find a root-hold on almost vertical slopes and the road is difficult as it winds on a raised path.

As the gradient lessens, at about 600ft, reindeer forest thickens, but at the top of the rise there is only bog and scattered trees. In summer, cotton grass grows around shallow lakes resting on peat. This is a foretaste of the tundra and sub-arctic conditions which characterise much of Lapland through which the rest of the Highway's path will lead. Before the road descends again, at 700ft, a further reminder that this is Lapland*: the summer huts of the reindeer Lapps who use this region to

---

*Throughout Finnmark (Norway's Lapland) the Lappish influence and presence is considerable. Although location references will be made in this chapter, a fuller account of the Highway and the Lapps will be found in Chapter 7.

pasture their herds. Most of this part of the Highway has now been raised and is kept open in winter.

To descend the 150ft to Leirbotnvatn, a large lake at the head of Lakselvdalen, the Highway follows the path of a stream, matching each shallow fall in the river with a steepening gradient. By the lakeside a minor road leads on down the valley to Leirbotn and Storekorsnes, leaving the Highway to climb gently again through Stokkedalen. This part of the road is narrow and especially dangerous in spring and even in summer after heavy rain. The surface is still only water-bound gravel, deeply gullied by rills. The valley is tree-filled and, in summer, the forests are full of reindeer which are liable to rush madly across the Highway as they wander in search of food. Below the road, to the south west, a river runs in a gorge which narrows towards the head, and near the top the forest thins as the plateau comes into sight.

The road levels at 1,250ft and prepares to cross the Senna vidda. This tundra wasteland is treeless, nothing but bare rock and bog. Sennalandet is desolate and uninhabited. The scenery is reminiscent of that across the Polar Circle on the Saltfjell. For miles the Highway runs across plateaux in what is really a broad and scarcely incised valley. The road, wisely, has been built above the valley bottom and the programme of raising was rewarded in the winter of 1969-70 when the road was kept open for the first time. The surface of water-bound gravel is adequate and, in many parts, the Highway travels absolutely straight for miles.

There are no permanent settlements on the plateau but, in summer, huts are occupied and tents erected by the Lapps who graze their reindeer here. Most of the huts are widely separated; at most, two or three families live within a few hundred yards of each other, but, near to where the Highway crosses the Repparfjordelv, a Lapp summer village has grown up at Aisaroiui (also: Aiseroaivve) (see Map 6). Complete with a new slaughter house and a church (*see plate, page 158*), the village is a real community which breaks the loneliness of the tundra. By the bridge over the river there is another reminder of wartime days. Behind the telegraph maintenance hut lies a huge pile of rusting food

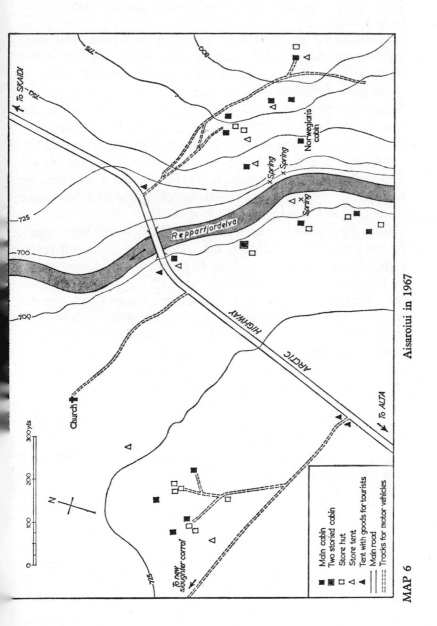

Aisaroiui in 1967

MAP 6

cans, the refuse of a German command post which occupied this
arctic desert. One can only guess how popular this station must
have been!

Beyond Aisaroiui, the Lapp encampments are fewer but, when
the Highway drops to 800ft and the birch wood reappears in the
shelter of the valley, an occasional summer chalet belonging to
Norwegians is tucked away among the trees. The valley of the
Repparfjordelv, here called Breidalen, narrows near Skaidi, the
river entrenched in a gorge within a gorge. At first the road follows the river, but at 600ft it levels out and drops less steeply,
eventually running on a narrow shelf over 200ft above the stream.
The gorge acts like a funnel to strengthen the winds blowing
through it. The trees are stunted, almost stump-like, but above
the gorge there is only rock and bog.

The gorge broadens into a valley at Skaidi with a more open,
forested landscape. Skaidi is not a village or even a hamlet;
rather it is a service point on the Highway. Sited where the road
to Hammerfest (Route 94) leads west off the Highway, Skaidi
has all the traveller needs: a café, a store, a petrol station and a
small guest house. The garage advertises *autoglass,* a reminder of
the hazards of driving on gravel roads, and a gravel works and
road maintenance depot are conveniently located nearby. Near
the road junction is a monument to one *Harald Hofseth, 1892-
1951, Vegsjef.* Born near Skaidi, Hofseth was chief road engineer,
responsible for planning much of the Highway through here.

The Highway turns sharply to the east at Skaidi after crossing
the Skaidelv (also: Skaidejokka), a tributary of the Repparfjordelv. The next section of the road demonstrates what can be
done with the expenditure of effort, time and money even this
far north. The Highway has been raised about ten to fifteen feet
above the old road level and its path realigned. The surface is
good oil-bound gravel and, apart from a little frost-heaving after
the spring thaw, is generally in excellent condition. The old
road, which was usually snow bound in winter and difficult in
summer, has been left and can be seen, sometimes to the left
and sometimes to the right of the new Highway, usually

Page 139 Hammerfest, the world's most northerly *city*. The main street is sandwiched between the bare plateau and the harbour.

Page 140 Honningsvåg at 71° north, typical of Finnmark's small towns and villages dependent for its livelihood upon the sea.

accompanied by the telegraph wires which have yet to be re-sited.

This part of the Highway's route is relatively simple. After crossing the Guraelv, the road leaves the forest to climb on to a low, bare and uninhabited plateau. This area is exceptionally cold and bleak in winter, for there is almost no protection from the arctic air. In summer, when mats of sub-arctic vegetation fill the gaps between the rocks and the narrow river valleys reveal their stunted birch, the landscape is a patchwork of grey, green and brown. The plateau's watershed is at only 700ft where the Hatter (or Haiter) ridges divide the drainage of Repparfjorden from that of the Porsangerfjord. From the divide, the Highway descends on a low gradient to Smørfjord Lake and then runs on an almost level course across the rocky fjeld towards Olderfjord. The old and new Highways merge before the steep descent of Olderfjorddalen, a tree-filled valley leading to the great Porsangerfjord.

Olderfjord is a hamlet at the head of a small island-blocked inlet on the western shore of Porsangerfjorden. Less than a mile to the north is Russenes, a larger village on the Honningsvåg-North Cape road (Route 95). This road follows the fjord shoreline, and so too does the Highway but in the opposite direction, after rounding the Kristrandfjell headland, its path is almost due south towards the head of the fjord. The headland is an ideal place from which to view the Porsangerfjord, the greatest of the northern fjords and, by area, the largest in Norway. Cluttered with islands and narrowed by headlands, this immense inlet forces the Highway into an expensive detour. Compared with a direct route which might be possible by ferry across the fjord, the Highway covers an extra sixty to seventy miles.

From Olderfjord to Lakselv at the head of Porsangerfjorden, a distance of forty miles, the Highway keeps close to the shoreline. Snow fences, some set two abreast, protect the road which, in recent years, has been straightened and its surface improved. The mountains bordering the fjord often rise sharply, leaving little room for the Highway, but at five promontaries the road leaves the water and crosses the neck of the headlands. These

J

rocky fingers protruding into the fjord are generally narrow and quite low-lying but, south of Indre Billefjord, a larger peninsula causes the Highway to rise 200ft over a forested saddle before rejoining the shore.

Many rivers flow under the Highway into the fjord and bridging has been expensive here (see also Chapter 2). Some of the old collapsed bridges, such as the old Salletjokka bru, have been left. They were even narrower than those on today's Highway and one wonders how any vehicle crossed them. Most of the rivers carry large loads of silt and gravel, but in the higher volumes and velocities of spring the quantity increases and boulders are a menace to the very structure of the bridges. The lighter gravels and sands spread into the fjord, as at the mouth of Stabburselva, or form terraces which can be exploited for road building materials. Many of the small gravel quarries border the Highway and, in places, the excavations have been limited by the road's accompanying telegraph lines. The telegraph poles often stand marooned on islands of gravel while mechanical excavators eat into the surrounding detritus.

The western shore of Porsangerfjorden is sparsely inhabited. The people here rely on fishing, which is less demanding on the limited space available, but almost always this is combined with the keeping of a few dairy cows. Along the Highway, at the entrances to the scattered farms, milk churns await collection on raised platforms, most of them covered to give protection from snow and rain. Standing like sentry boxes at the farm gates, these loading platforms, painted green or red and white, have a letter box attached, a reminder of the Highway's importance in carrying mail to these isolated communities.

There are a few villages and hamlets between Olderfjord and Lakselv. Kistrand is the first. Formerly important as a port of call for coastal steamers, Kistrand remains dependent on the sea for its fishing industry. Fish-drying racks and nets litter the beach and a small meteorological station hints at the control which the weather must exercise on the villagers' daily lives and livelihood. There is a pleasant little church in the village and, as also further up the fjord, signs of the German wartime occupation are evident.

South of Kistrand, some fifteen miles away, is the hamlet of Indre Billefjord situated in a picturesque cove, sheltered to north and south by peninsulas. The southern headland is broad and a number of fisherman-farmers have settled here, particularly on the southern side which combines lower slopes with a more favourable aspect.

The use of Finnish instead of, or in addition to, Norwegian becomes increasingly in evidence along the Porsangerfjord. In part the explanation lies in the origins of many of the families that migrated to this Finnmark kommune in the eighteenth and nineteenth centuries, but the Finnish café and lodging signs also reflect the number of Finns who come into this area in summer via the old Karasjok route. The Finns are notoriously uni-lingual.

Near Øvrenes is the Stabburselv valley which claims, probably accurately, to contain the world's most northerly coniferous forest. Stabbursdalen has other attractions, such as Njakkafoss, a waterfall running through a narrow gorge, and, for the sportsman, elk and salmon. It is another of those valleys only touched by the Highway which invite closer inspection. From the hamlet of Øvrenes the great Finnmark vidda dominates the view south; westward the valleys are broad, their concave slopes scoured by ice.

On the last part of the Highway into Lakselv there is a striking contrast between steep and impressive rock faces—rising 1,000ft and scarred by weathered clefts and erosion gullies—which lie on one side of the road, and the alluvial flats on the eastern or fjord side. The bays show all the signs of natural reclamation and the large sand banks of Vesterbotn and Brennelv expose the shallowness of the fjord head at low tide.

Near the entrance to Lakselv, from the north, is a military area and the Highway enters the village at Skallenes where, in pre-war times, one had to use a ferry to cross Lakselva. The settlement is, if anything, even more dispersed than at Alta, but it lacks the latter's charm and, although smaller, is definitely more urban in appearance. Strictly, there are a number of hamlets which make up an agglomeration of about 1,000 people, Banak and Brennelv now being part of Lakselv. All the buildings are very new: the church, a block of flats, the bus station, modern houses and stores. The hotel is probably the least attractive of all those on the Highway, but is often full in summer because of the importance of communications to the village and the river's reputation among anglers (*Lakselv* means salmon river). Lakselv, at the junction of the Arctic Highway and the Norway-Finland Route 96 through Karasjok, is one of the overnight stopping places for the North Norway Bus. There is a civil and military airport at Banak. The runway, like that at Alta, is built out on the delta flats of a river, here provided by Lakeselva.

Outside Lakselv the Highway continues to skirt Porsangerfjorden across the low-lying head. Around Øvre Brennelv there is some farming beyond Karijokka bru as far as Časkel, where there is a moderate sized timber yard. Farming is by no means easy. The soils are sandy and sharply cut by streams which, in spring, carry away their own banks. Where unworked, the sandy flats are covered with masses of cotton grass in summer.

After crossing Časkeljokka bru, the Highway follows the edge of the fjord closely for about twenty miles through totally uninhabited territory. The surface is still poor but the road is wide enough. The view into the fjord is quite wonderful. The silver-grey dolomite cliffs of Reinøya are seen at their best in the light of a low sun.

Inland from the road, the mountains of Bjørnefjellet, Munkkovarre and Tverrfjellet rise to over 1,500 feet close to the coast. Rugged and steep, their bare slopes are exposed like a section on a geological map. The torrential rivers which drain these mountains are arrested as they meet the still water of the fjord. Moraines which block the river mouths are being eroded by the streams making gorge-like incisions into the gravels and carrying boulders on to the beach. Boulders are strewn also against the fjord side on both margins of the Highway by the weathering of the rock faces. Avalanches are quite common in this area and the road is often affected by heaving in spring.

Towards Børselv there are prominent raised beaches and, after crossing the river Capper, the mountain wall moves back from the Highway to give way to a hummocky bog with some birch.

At Bibaktad there is a small farming-fishing hamlet lying in the partial shelter of the Hestnes point. Two miles further on, a path leads down to another hamlet, Vækker (also known as Surrukopp), and the Highway then leaves the fjord to its crossing of Børselva. Most of the settlement lies to the west of the bridge, off the Highway, and the community farms the alluvial soils at the river's mouth. The village has a quite large boarding school, attended by Norwegians and Lapps, and a notable church. A central spire crowns the wooden church which is in the general shape of an irregular octagon with two lateral extensions giving it bi-axial symmetry. Many of the villagers are of Finnish stock.

To cross from the Porsangerfjord to Laksefjorden the Highway strikes an easterly path to cut off the dividing peninsula. The route is not easy; it demands a climb to over 600ft across an undulating plateau which, for thirty miles, is without habitation. The difficulties of this section of the Highway are readily seen by its long curving path, and winter closure is unavoidable. The need for an all-weather road here is, however, not pressing and the expense, even of raising the road, would be difficult to justify.

On leaving Børselv, the Highway follows the broad river valley which is farmed and forested. Then, gradually at first but later more steeply, the road climbs on a winding path up out of the valley and on to the plateau. Birch predominates only to thin and finally to disappear at about 500ft. The Highway parts company with the river on leaving the village, but on the path to the plateau Børselva is often only a few yards from the road. A few yards but hundreds of feet, for the river flows through a truly magnificent dolomite gorge which is almost totally hidden from the Highway by birch trees. Looking down from the edge of the gorge, the crystal waters of the Børs river can be seen tumbling over small falls and into silent pools. The river is rich in salmon and trout but, in late summer, the tiny speck one can see at the bottom of the gorge is probably an angler seeking the red char for

which the Børselv is famous. Up stream from the gorge there are
deeper falls, the Silfarfoss and Meksafossen, beyond which the
river is incised into glacial drift.

The lower part of the plateau is covered with short trees.
Reindeer fences, the occasional stockade and rising smoke
identify this in summer as a Lapp region. In fact the whole of
the plateau is used for grazing reindeer in the snow-free months.
Børselvfjellet, for the Karasjok Lapps, is what Sennalandet is for
the Kautokeino Lapps.

As the altitude increases, the stunted birch thins and is finally
replaced by undulating, hummocky bog land splashed by small
lakes. In parts there is practically no vegetation at all. Instead,
ice-scratched crags and the powdered rock dumped by dead
plateau ice occupy the surface. Hollows are filled by lakes or, if
sufficiently deep, by dwarf birch. This is indisputable tundra:
bleak but beautiful.

The water divide on Børselvfjellet is 625ft high and, after
crossing it, the Highway follows the line of the Luobbaljokka and
then Storelva. Further reindeer stockades, a new large store house
and a Lapp chapel, hidden among trees, stand close to the road.
The Highway's gradient, as it descends the eastern side of the
plateau, is gentle, but the road surface is poor and the width
accommodates only a single vehicle for the most part. The landscape is one of lakes and streams, birch and bare rock. Towards
the bottom of the descent the Storelv broadens and, in low water,
its course is braided.

The Highway returns to sea level at the hamlet of Storfjordbotn, which lies at the head of the Storfjord arm of Laksefjorden.
Ringed by black frost-shattered mountains, an aged, weathered
landscape, the hamlet is terribly isolated and somewhat dingy in
appearance. Snow scooters, stored in the open in summer, are a
reminder that roads are only seasonally significant here.

Just outside the hamlet, the Highway passes weather-sculptured
rocks to climb slightly away from the fjord through a forested
valley. A drop again brings the road across another pass,
absolutely flat, with shallow sides, and over the twin bridges

of Evafossen and Adamfossen. The 1961 bridges cross two parts of the same waterfall, the Adamfoss, which drops just over 120ft from the Adamsdal, which hangs in respect of the fjord.

Only a fleeting glimpse of the Laksefjord is possible from the Highway. An almost hemispherical little island lies in a nearby bay and there is a special quietness and serenity about the whole of the fjord. The scenery more than compensates for the poor surface of the road as it leaves the fjord to slice across another peninsular neck. The landscape is stark but not harsh, untamed but not menacing. Its primeval quality is its fascination. There is a unique beauty about bare rock when, after being washed by rain, it is caught glinting in the sun. Its grey face is transformed to a surface of newly polished pewter.

A gorge beyond Frierfjorden is eerie but the plateau to which it leads is open and windswept. Great roches moutonnées separate shallow lakes, and glacial erratics lie abandoned on spurs and hilltops like the playthings of a mountain giant.

Where there is sufficient shelter, small trees grow. Elsewhere, bog grasses and mosses search out every likely crevice to obtain a root-hold. The land offers little to man, who prefers instead to exploit the riches of the fjord. In fact, few people live in this part of Finnmark and improvements on the Highway, its straightening and raising, are justified only by non-local traffic.

A delightful little cove at Landersfjorden shelters a small hamlet; about half a dozen families live in each of the hamlets around the more open Frierfjord. But, apart from work at a road maintenance station at Solbakken, these people are almost totally dependent upon fishing.

The Highway leaves Laksefjorden after grazing the tail end of the inlet at Ifjord. The village here is rather larger than others on Laksefjorden. Apart from the usual collection of fishermen's houses there is a café and a petrol station. The problems of the road's next section are predicted by the presence of a road maintenance depot and a control point of the NAF (Norges Automobil-Forbund). The next section is, in fact, the Ifjordfjell.

In crossing from Laksefjorden to Tanafjorden, the road has to rise to over 1,200ft to find a path across Ifjordfjellet; yet it is never more than five miles south of latitude 70° 30′ N. Not surprisingly, this part of the road is closed in winter, usually for longer than any other portion of the Highway. It is a good year that sees the road in use earlier than June or for more than six months. In summer the gravel surface is often quite appalling, its already narrow width further reduced by collapse at the edges. Vehicle passing places, marked with the sign 'M', jut out dangerously to overhang abrupt slopes. In all seasons, tall birch sticks mark the edge of the Highway but are eventually submerged by all-enveloping snow.

To reach the top of the plateau, the road first climbs sharply up the valley of Ifjordelva, then passes over a col at about 800ft before bridging a stream which links the Iskløver lakes. Even at this height the trees have long since been left behind. In mid-summer patches of snow still lie well below the level of the road, and snow may fall in any month.

At a little over 800ft there is a small Lapp encampment. This is the nearest approach to a Lapp summer *village* that I have come across, saving that on the Senna plateau. The large corral is usually littered with skulls and bones, as well as pieces of reindeer skin. Unlike Aisaroiui, it is an untidy community, but the new wooden chapel, resembling a tent and on rising ground just south of the Highway, is enchanting.

Needless to say, there are no permanent settlements on Ifjordfjellet. The Lapps' use of the plateau is shown by the reindeer fences that cross even the summit of the fjeld and, come rain or snow, anglers find sport in the lakes and streams which cover so much of the surface. But, in winter, none will venture on to this arctic wilderness.

The Highway crosses the water divide at 1,200ft, rather less than halfway across the plateau. The landscape is roughly hewn; monotonous and bare, yet always interesting (*see plate, page 70*). Many of the lakes and snow patches are large in area but shallow. For the most part, the rivers run in broad valleys of little depth but, where small vigorous streams flow on steeper slopes or a mass of glacial drift lies on the surface, then the rivers are incised into narrow V-shaped cuts. There is no shortage of water in summer, as the streams feed on the slowly melting snow, but in winter there are no streams, no lakes, no rock, just a thick white cloak to give a sameness and a silence to the plateau.

To descend from Ifjordfjellet, the Highway begins by using the same path as the Storelv, but there is no clear natural route. After following a narrow valley, the road breaks away from the river. The river passes through a gorge, while the road winds its way down across a terrain so broken and irregular that it defies description. Suddenly the Vestertana arm of Tanafjorden comes into view: a glimpse at first, then fine views of the glistening inlet. But this proves to be only a teasing glance, for the Highway cannot easily leave the plateau. Only after following a tortuous path, full of hairpin bends, does the road pass round another gorge and finally reach the fjord. Fortunately, the ledge, which is the Highway's track, has been widened and improved in recent years; previously the descent was perilous.

Vestertana is shallow and sand-filled. It is one of the many inlets which dissect the head of Tanafjorden west of the river mouth and cause the Highway to rise and fall over the hilly peninsulas which separate them. Although each inlet is not especially long, the road is forced to climb over 400ft between Vestertana and the little Tarmfjord and again to reach Smalfjorden from Tarmfjorden. Finally, a lower divide of 200ft is used by the Highway to join the Tana river at Rustefjelbma.

Each small fjord has some settlement. Between Sjursjok in Vestertana and the village of Smalfjorden there are small hamlets and isolated homesteads making a living from river and fjord fishing and from the limited amount of farming that can be practised this far north. Torhop and Smalfjorden have quite sizable jetties and oil storage tanks and there is a small school at Torhop. Fish are left to dry on racks set up on the beaches in summer while the fishing boats explore the Tanafjord for another catch. The vessels used are incredibly small: many are scarcely forty feet long and few exceed fifty feet. With their box-like cabins, they appear almost comic and certainly unseaworthy. That they set out in almost any weather and invariably return laden is a tribute to the skill of both fisherman and boat builder.

The hilly divides between the inlets are bleak and treeless, but the lower slopes, which slip gently into the shallow fjords, are clothed in birch which is remarkably thick for this latitude. It is just beyond Torhop, where the road turns towards Smalfjorden, that the Highway reaches its most northerly point, coming to within a mile and a half of latitude 70° 30′ N. From here its path to Kirkenes is south-easterly.

Each time the road touches the head of an inlet there are splendid views across Vestertana towards the mountain peaks of Flattind and Perletind. These mountains rise to well over 2,000ft and are part of the peninsular divide between Vestertana and Langfjorden. The slopes of this peninsula are particularly steep on the south-east facing side—2,000ft is reached only two miles from the shore—and silvery streams make their descent by a succession of falls over bare rock.

The Tana is variously described as the longest, second longest or third longest river in Norway! But its true position in the river hierarchy is of no consequence; it is certainly one of the greatest and perhaps the most majestic in the country. The Highway reaches the river at Rustefjelbma, a Lappish name meaning 'the cove of the rusty coloured water'. There is no village as such here; instead settlement is scattered along the terraces which form the banks on both sides of the river. Although Tana is often mentioned as though it referred to a single village it is really the district name.

Close to the point where the Highway and river meet is the new Tana church. Built to the design of Esben Poulsson, the church was officially opened on 21 June 1964. It is a rather plain wooden building with a steeply canted roof, but it has a most attractive tall steeple. This pyramid of wood, with a belfry, stands away from the main building to which it is connected by a corridor. The whole structure is very modern and in singular contrast to the old Tana church at Langnes.

A minor road, in very poor condition, joins the Arctic Highway at Rustefjelbma. This track—for it is scarcely more than that—runs along the left bank of the Tana and out to the mouth at Nordre. Along the road, at Bonakas, is the world's most northerly agricultural college, as well as many older buildings including the Langnes church. At Nordre there is an unobstructed view along the whole of the outer Tanafjord and into the Arctic Ocean.

But the Highway sees none of this, for at Rustefjelbma it turns sharply southwards following the river up stream. Many of the great terraces which the river has formed are farmed, but the farms are separated by tracts of forest. Tanaelva and its valley are quite exceptionally broad, the mountains standing back a respectful distance from the river. For miles the Highway travels over a bumpy but otherwise gradient-less course. Even the bridges across tributaries, like the Maskejokka which has incised spectacular meanders into the terrace, fail to cause the Highway to deviate from an almost perfectly straight path. The farms on this left bank are matched by others on the opposite side of the river where the parallel terrace carries another road. One farm, near Rustefjelbma, has a quite sophisticated meteorological station.

Twelve miles south of Rustefjelbma is Vestre Sieda. Here the old ferry used to cross the river and in wartime the Germans erected a pontoon bridge. Now the Highway continues further south for a few miles, swinging away from and losing sight of the river. A series of terraces lines the river, the lower ones farmed and the upper terraces used for pasture or, more often, forested. The road stays with the lowest terrace until, at a narrow bend in the river, it makes its crossing over the elegant Tana bru. This suspension bridge has a span of nearly 600ft.

From the bridge something of the great width of the river can be appreciated, flowing between the cliff-like terraces it has carved. Massive sand banks lie exposed when the river is low and salmon of enormous proportions swim up stream towards the Stor falls and Aile rapids, well known to anglers.

The Arctic Highway turns south again after crossing Tanaelva and continues up stream to the hamlet of Skipagurra where salmon are landed prior to export. A narrow crumbling road leads off the Highway from Skipagurra to Polmak near the Finnish border; in winter it is possible to drive on the frozen river. Polmak is the smallest of the three Lapp villages in Finnmark, but in some ways it is the most interesting. More compact than Kautokeino and without the influence of tourism that spoils Karasjok, Polmak's treasure is its church. Built over a century ago, the small wooden church has an altar panel dating from 1625. The village achieved a special fame in 1928 when its postmaster Henriksen landed an 80lb salmon to create a record for the Tana river which I believe still stands. The village does not have an exclusively Lapp population. Norwegians have settled here since 1790.

To cross from the south-north Tana river to the east-west Varangerfjord, the Highway turns inland from Skipagurra. Rising away from the river, the road passes through forest on to the boggy and lake-splattered Seidafjell at over 400ft. Much work has been done in recent years to improve this five-mile section of road. By raising the road, sometimes forsaking the old path, and reducing the curvature of the bends, the Highway is beginning to be recognised as an important link between two communities. The Tana river valley and the Varangerfjord are, by far, the two most settled areas in the whole of East Finnmark.

The Seidafjell is important too for the Lapps, but their movement, in their bi-annual migrations, is across the Highway rather than along it. For centuries the Polmak Lapps have moved over Seidafjellet with their reindeer to their summer pastures on the vast Varanger peninsula on the north side of the fjord. In the past, reindeer were hunted and trapped in this area. A complex of ditches and pits were dug and stone walls erected to capture wild reindeer. Some pits remain near the Highway on Seidafjell, but the largest system of over 500 is on the fjeld to the south. Today, the Lapps drive their semi-domesticated herds across the Highway in May and October. Perhaps 6-7,000 deer cross the Seidafjell.

On the plateau the road rides high and there is little to obstruct the view. To the north and south is open and uninhabited tundra but to the east is Norway's *last* fjord: Varanger, over sixty miles long and the only major fjord in the country to have its head in the west and its mouth in the east. A low peninsula extends into the fjord to divide its head into two arms: Meskfjord and Karlebotn. It is towards the former that the Highway makes a moderately steep descent from the Seida plateau, reaching the fjord at Varangerbotn. The Highway then turns to follow the southern shore of Varangerfjorden, leaving a far superior road, Route 98, to serve the fylke capital, Vadsø, on the northern coast.

Although, as in all Finnmark, most of the Varangerbotn buildings are new, Lapp influence can still be seen. Farms occupying the plains at the head of the fjord often include turf store houses, some being of the same construction as the gamme. There is a small hydro-electric station at Varangerbotn, but it is after crossing the Vesterelv that some of the newest buildings can be seen at Karlebotn. These include the Karlebotn school and the Nesseby Ungdomsskole (or youth school). The latter, a State boarding school drawing its pupils from all over East Finnmark, not just from the kommune of Nesseby, is pleasantly sited near to the fjord.

Karlebotn is of considerable historical interest. Almost continuous settlement can be traced here from the stone-age Komsa culture. The Gropengbakken site has yielded a number of finds, especially in the middle beach terrace where nearly ninety separate sub-sites have been located. The early Lapps hunted seal and trapped whales in the shallow waters of the inlet, but their descendants have settled by the shore and are content with fishing.

Up to the nineteenth century, Karlebotn was the main trading post for East Finnmark. It had its own courthouse and prison and an annual fair was held. Now, much improved road communications along the fjord relegate Karlebotn to the status of a small village, Vadsø has assumed its rôle as administrative centre.

On leaving Karlebotn, the Arctic Highway sets out on the final part of its journey to Kirkenes. This last section of road is, perhaps, the poorest in terms of physical condition and is the least used of any major segment of the Highway. Obviously, the exceptionally long and severe winters are contributory to this state of affairs, but equally important is the fact that the settlement along this route still looks to the sea as its primary means of communication. After all, the sea is available for twelve months of the year, it is the source of the people's livelihood and the hamlets are at the coast, while much of the Highway is built inland.

The southern side of Varangerfjorden is bleak and without the farms that fringe much of the north coast. Aspect plays its part, but so too does the more hummocky rock surface over which the Highway passes en route for Kirkenes. Inland from the road the landscape is strangely unearthly. Scarred rock surfaces, weathered and without vegetation, streams, more gentle than before, flowing from black mountains: all is somehow unwelcoming.

Yet if one looks seaward from the road, from those points where the fjord comes into view, then the contrast is manifest. Illuminated by the weak arctic sun, the other side of the fjord is altogether different. Here the greens and browns of vegetation form a background for glistening pink-white and grey-white rock scars. Behind, the mountains rise to snow flecked peaks and, in front, the northern fjord coast is lined by a scattering of brightly painted farms. Only in winter is the contrast smothered by snow and obscured by enveloping darkness.

In the fjord, small islands of rock are crowned with white painted warning lights and in the bays on the southern shore lie fishing hamlets with a mere handful of small farms. These hamlets enjoy sites of unforgettable beauty. Sheltered within the bays, small boats bob up and down on blue-green waters. Red and yellow, white and blue wooden houses stand on the shore. Fish and hay are out to dry on the beaches or in the Lilliputian fields. Again, only in winter is the harsh reality of their situation evident.

The two most delightful hamlets passed by the Highway are

Grasbakken in Veinesbukta and Gandvik in Gandvika. Grasbakken has a substantial jetty and its farming includes a flock of sheep. Gandvik, exceptionally well sheltered in a mountain-girt cove, has a small hydro-electric plant fed by water channelled down almost vertical pipes off the mountains.

Much of the population of East Finnmark is not Norwegian but Lapp or Finnish—or even Russian. Just by Gandvik, a narrow local road leads off the Highway to Bugøynes at the tip of the peninsula formed by the Bugøynesfjell. This village is more Finnish than Norwegian. The names in the cemetery, the language in the streets, the style of the older buildings—all are Finnish. Indeed, there are some old buildings here, for the Germans did not destroy Bugøynes as they did the other villages in the withdrawal of 1944. Whether this was because of its Finnish connections or its geographical isolation is not clear. Today, new buildings outnumber the old and, with a prospect to the mouth of the fjord, its inhabitants are largely fishermen.

From Gandvik the Highway turns inland, never to return to the open fjord. The road climbs on a steep gradient to a hilly pass across the neck of the Bugøynes peninsula. At its highest point, over 500ft and momentarily out of the pass, there is a fleeting view of the fjord and its shallow bay-head beaches. Then again all is desolate: lakes, frost-shattered rocks, valleys choked with rock debris.

After crossing into the kommune of Sør Varanger, marked by a sign of welcome in four languages, the Highway runs alongside Hauksjøen lake, sinuous and beautiful. The lake rests on the plateau and drains south-eastward towards the Bugøyfjord. The Highway follows the Hauksjøen valley, which widens and fills with birch. After crossing the river, the road passes through Sopnes, where a track leads to the hamlet of Valen, and then immediately descends into the village of Bugøyfjord.

This is another of those international villages, like Skibotn and Bossekop, which attracted Finns, Norwegians and Lapps to its market; here, too, came Russians. A special importance was attached to Bugøyfjord, for it was here that the old winter snow road, from Inari in Finland, reached the sea.

There is a large road maintenance depot in the village, but otherwise it is now indistinguishable from the other fjord villages. The main Varangerfjord cannot be seen from Bugøyfjord; in fact the inlet has such a narrow exit that, from the village, it appears to be enclosed like a mountain lake.

The southern perimeter of the Bugøyfjord is well farmed as far as Vagge but, as the Highway turns away from the fjord, it rises up the wide valley of Klokkerelva which forms the road's path over the plateau separating Bugøyfjorden from the Neiden-fjord. The plateau is ill-drained; two great marshes, Førdes-myrene and Sakrismyrene, cover the surface and the Highway climbs to nearly 500ft. With so much surplus water about, the problems of frost heaving are aggravated and this section of road is troublesome in spring.

As the Highway swings round the mountainous Norskelvfjell, it drops from the plateau to the valley of Neidenelva. This river rises across the border in Finland, but in its lower, Norwegian, course it provides a wide and fertile valley which is carefully cultivated and has attracted settlement for centuries. The settlement here is effectively divided into two by a waterfall: Skolte-fossen. The Highway crosses the river above the falls at Neiden and below the road bridge salmon are caught by rod and by net (*see plate, page 88*). Above the fall is Neiden, a small village which in winter is connected by snowmobile with Route 4, the Finnish road to Ivalo. Downstream is Skoltebyen.

Skoltebyen (literally: *the town of the Skolts* or Skolt Lapps, but, in fact, a village) is of very considerable interest as the only Skolt Lapp community left in Norway. Close to the Highway, on the northern side of the road, is the community's Orthodox Chapel, unique in Norway. This tiny log chapel is only a few feet square, with space for perhaps six or eight people. Incongruously, a modern and expensive-looking record player stands at the opposite end to the altar. The key for the chapel is held at the nearby farm bordering the road.

The lower part of the river Neiden is deeply incised between

Page 157 Kvænangsfjellet, a Lapp's turf hut (*gamme*), rarely inhabited now but still something of a tourist lure. Note the two types of *gamme*, rounded (typically northern) and pointed (southern), and the snow fences.

Page 158 (*above*) Lapps leaving the chapel at Aisaroiui. Most Lapps keep their best clothes for just such an occasion. (*below*) Lapp woman in summer cabin, making reindeer skin boots for sale to travellers on the Highway. The old iron stove is typical but the decorated chest seen on the left is rapidly becoming rare.

sand and gravel terraces as it meanders towards the fjord. But the Highway moves away from the river mouth and over a low marshy depression to the head of Munkfjorden, an arm of the Neidenfjord. Once the Munkelv is bridged, the road swings towards the steep edge of the inlet and follows an undulating path along its shore. Two or three houses are built on each of the flats that protrude into the fjord and the views from the road are pleasing if not spectacular.

The eastern shoreline of Neidenfjorden, which is followed by the Highway, is broken by a narrow strait linking this fjord with Korsfjorden. The whole of this part of the Sør Varanger coast is fragmented by inter-connecting fjords separated by rocky peninsulas and islands. Something of the coast can be seen from the Highway, but the open Varangerfjord is never in view. From Tusenvik, the road cuts across the marshy neck of Tømmernes and rises to over 300ft.

Kirkenes is now only a few miles away; so too is the Soviet border. First, however, the Highway passes through a military area with its prohibitions: 'no photographs, no camping', and its warning that it is dangerous to move off the road. Ironically, in this same area, by the side of the Highway, is a war memorial to the Russians buried here in the last war.

The Highway is asphalted near the army camp and over the last miles into the town. A road to the civil and military airport leads off at Høybuktmoen and the Highway then descends towards Langfjorden. The river, which links the fjord head with Langfjordvatnet—a narrow lake nearly twenty miles long—is crossed by the new Straumen bridge. A short climb over a rocky ridge and the Highway enters Hesseng, a sort of suburb to Kirkenes. Here there is the junction with Route 885, the road which leads to the Bjørnevatn iron ore mines and which, during the war, was linked to the old Finnish Arctic Road.

The Highway enters Kirkenes through a residential district and alongside the railway bringing ore to the port. The Arctic Highway has reached its destination. At latitude 69° 45′ N and longitude 30° 10′ E it is as far north as central Greenland and as

K

far east as Istanbul or Alexandria. A journey of over 900 miles has brought the Highway from Mo i Rana, and all but a small fraction of that distance has been inside the Polar Circle.

Kirkenes owes its present importance to iron ore. In 1900 there were only four or five houses on the headland site which extends into Bøkfjorden between the Langford and Glvenesfjorden. The re-discovery, in 1902, of sedimentary iron deposits just seven miles to the south at Bjørnevatn (Bear Lake) led to the establishment of a mining company, Aktieselskabet Sydvaranger, in 1906 and to an influx of people from south and central Norway, as well as from Finland and Sweden.

The first low phosphorus high-grade iron ore concentrate was produced in 1910, but the two world wars brought crises and difficulties which extended into the post-war years. Early capital came from Sweden and Germany, but the reconstruction in 1948-52 was helped by State loans and Marshall Aid. The forty-three per cent pre-war German shareholding became the property of the Norwegian Government in 1945, increasing their holding to fifty-one per cent.

Diesel-electric locomotives haul the ore in eighteen-truck loads to the crusher in Kirkenes, where it is briquetted or pelletized for shipment. Over a million tons each of concentrate and pellets are exported annually to USA, the UK, West Germany and Finland, as well as to the Norsk Jernverk plant in Mo i Rana. Dredging in 1967-8 increased the capacity of the port to accommodate ore-carriers up to 80,000 tons, and loading rates of 4,000 tons per hour can be achieved.

The population of Kirkenes today is close to 5,000, with a further 2,000 in the mining settlement. A/S Sydvaranger has become the largest mining concern in Norway.

Kirkenes has been rebuilt since the war. It suffered the fate of most of Finnmark when it was burnt to the ground in the German retreat of 1944, but much had already been laid waste in four years of air raids. It is said that only Malta was subjected to more continuous air raids than Kirkenes in the whole history of World

War II. As the war drew to a close, the fortified town lay in ruins from the summer raids of 1944 and the population was forced to live in the mines. As the Russian army drove out the Germans in October 1944, over 3,500 people sheltered in the mines and no less than ten children had the Sydvaranger mines recorded as their birth place.

Russians stayed in the area for nearly a year before they withdrew and reconstruction began. On the western side of the town are the ore-crushing works leading directly to the ore quays. The central and eastern part of the town is laid out on a grid-iron plan with multi-coloured wooden houses and more sombre office blocks. The gardens and grass verges to the roads make this as pleasant a town as any in North Norway. There is a general air of prosperity and sophistication which contrasts Kirkenes with most arctic settlements.

Nearly 1,000 are employed by the ore company and there is little diversification of occupations. A saw mill processes timber from the Pasvikdalen *taiga* forests and a plastics factory—a joint Finnish-Norwegian project—has been planned; but, as with Narvik, iron ore continues to be the town's *raison d'être*. Attempts to attract tourists have met with only limited success. True, there is a very comfortable hotel sited close to the Highway's entrance to the town, but its guests are more likely to be businessmen visiting the mines or ore works than tourists.

In its isolated situation in north-east Norway, with only the USSR to the east, Kirkenes is the terminal of a number of major communication lines other than the Arctic Highway and the North Norway Bus. The port is the terminus for the *Hurtigrute*, the express coastal steamers, and Høybuktmoen airport is the final destination of aircraft on the SAS west coast flight from Oslo and of aircraft from Helsinki. In fact air and sea continue to be the most important links that Kirkenes has with the outside world. It is not to diminish the status of the Arctic Highway to admit that with its closures, poor surfaces and, above all, its immense length, it faces strong competition from the air and the sea when it comes to long-distance travel.

Chapter Six

The Highway's Branches

The Arctic Highway is a trunk road. In its passage through North Norway it bypasses many of the coastal and island communities as well as the more remote inland villages of Finnmark. Today, nearly all the larger settlements are linked by branch roads to the Highway such that it has become the distributor road for the whole of the three northern fylker.

It is beyond the scope of this book to describe in detail all the branch routes or to fully discuss their significance. However, much traffic uses the Highway because of the settlements or other attractions which lie along the secondary roads. The fruit, as it were, is at the end of the branches.

Few travellers along the Arctic Highway resist the temptation to explore at least some of these branches, for it should not be assumed that they are inferior to the Highway. In fact, many are wider, some are totally asphalted and most are well maintained. The difference between State roads (*Riksvei*), County roads (*Fylkesvei*) and Kommune roads is basically one of responsibility for maintenance and construction.

There follows a complete list of all the Arctic Highway's more important branch roads, together with a brief description of these secondary routes and some of their attractions. As with the Highway itself, the branch roads are listed in sequence from south to north, from Mo i Rana to Kirkenes. The Route numbers are those of the Norwegian Highway authorities and incorporated on signposts. The list numbers, 1-33, are used on the location maps, Maps 2, 4 and 5.

1. Route 77 : The Butter Road : Mo i Rana to the Swedish frontier

This road leads off the Highway just north of the centre of Mo. Completed during the war, it was formerly only a track through the mountains to Umbukta. There has been much improvement recently to the surface, width and cut-offs, but away from Mo, on the steepest section, there are still some dangerously narrow hairpin bends. Route 77 is subject to closure in winter, but in summer it provides, at present, the Arctic Highway's only direct link with Sweden north of Mo i Rana. There are some exceptionally fine views from the mountain summit, and the Swedish continuation road (Swedish Route 361) is first-rate and full of interest. (See also Chapter 3.)

Distances: Mo i Rana to Swedish frontier: 25 miles/40km; to Stockholm: 680 miles/1,095km.

2. Route 805 : The Blue Highway : Mo i Rana to Nesna

The Viking village of Nesna lies at the mouth of the Ranafjord on the northern side. The junction of Route 805 with the Highway is at Selfors bridge where the latter leaves the town for the North. This branch road, beyond Mo, is gravel surfaced and only moderately well maintained, but generally level. Although passing through a number of old settlements, the road itself is quite recent. As well as providing a useful link with the open coast, the route affords some good viewpoints into Ranafjorden and, especially after crossing the Sjoneidet isthmus, into Sjonafjorden.

Distances: Mo i Rana to Nesna: 44 miles/70km.

3. Minor Road : Røsvoll towards Svartisen

At Røsvoll, seven miles north of Mo i Rana, a minor road leads from the Highway westward towards Norway's second largest glacier: Svartisen. Most of this local road is extremely narrow and poorly surfaced with water-bound gravel, but some limited improvements are likely in the future because Mo's airport is situated about one mile west. The simple landing strip, hidden among trees, is almost invisible from the road.

Apart from the beauty of the glaciated valleys with their almost vertical walls and milky rivers, the route offers two special attractions : the Grønli caves and Svartisen glacier.

The Grønli caves have none of the commercialisation which spoils many similar limestone phenomena in Western Europe. The entrance to the caves is by way of a steep and narrow path leading off the road. A nearby farmhouse will supply a guide and the caverns and subterranean rivers are of considerable size and interest. Apart from the expected labyrinth of caves, there are some good examples of dripstone formations which the imaginative see as the pillars of a church; they have been called *Storkirken*.

The plateau glacier of Svartisen (Black Ice) is reached by way of Røsvassdalen, Svartisdalen and Lake Svartis. The road stops at the south-eastern end of the lake and, in summer, a small motor boat, *Svartis II*, carries passengers the final two and a half miles.

Svartisen covers over 200 square miles but is split into two by a deep valley, Glomdalen. On either side of the valley, two major ice caps, Østisen and Vestisen, occupy the plateau as great arched shields of ice. Though now classified as *dead-ice*, these caps have a recorded history of change. Their present size is similar to that of three hundred years ago, but at the beginning of the eighteenth century the glacier began to grow. Its northerly valley tongue, Engabreen, pushed out towards the sea in Holandsfjorden, swamping two farms in its path. The valley glacier, Østerdalsisen, slipped southwards into Svartisvatnet. Such was the position fifty or sixty years ago, since when the outlet glaciers have been in rapid retreat. Østerdalsisen has melted back about half a mile in the last quarter century.

A dangerous situation developed at the beginning of the war when, for some reason not fully understood, Østerdalsisen failed to discharge most of its summer meltwater to the west. Instead, a large lake formed at the outlet into which the glacier calved.

The lake itself began to discharge eastward, under the glacier, to flood Røsvassdalen. The remedy was a mile-and-a-half-long tunnel which relieved the pressure and made safe the scattered valley farms.

To see the whole of Svartisen it is necessary to fly over the ice cap through which *nunatakker* (Eskimo: lonely peaks) rise to well over 5,000ft. What with these peaks and the cold surface at times producing thick clouds, flying conditions are often difficult. It is certainly easier to use the road and lake.

On landing from the motor boat at the west end of Svatisvatnet, there is a long climb over rocks scarred and shattered by ice, but the rewards more than compensate for the effort: ice-blue caves, water- and ice-falls, the ice and snow of Østerdalsisen, all in an unforgettable setting.

For the glaciologist, geormorphologist or ecologist, Svartisen is a field laboratory. It is rare that there is not at least one scientific party, usually from England, camped out by the edge of the glacier in summer.

The road to Svartisen is subject to closure in winter and is very poor in spring. *Svartis II* is not available except in the summer season, although it is possible to walk from the end of the road, round the lake and up to the Østerdalsisen glacier.

Distances: Røsvoll to Grønli: 6 miles/11km; to Svartisvatnet: 15 miles/24km.

4. Route 80 : The Maelstrom Road : Fauske to Bodø

Leaving the Arctic Highway to turn northwards out of Fauske, the road to Bodø continues westward along the north shore of Saltfjorden or the Skjerstadfjord. This is an excellent road, open throughout the year and almost without gradient. It has an asphalt surface except for some short stretches. Because it links the Nordland capital with Fauske and the Arctic Highway, traffic tends to be relatively heavy, despite competition from the Nordland Railway which, except between Røvik and Straumsnes, follows an identical route.

From the neck of the Fauske peninsula, which is the town's

site, the road at first closely follows the edge of Klungsetvika and Skjerstadfjorden. The shore is steep and the road has to run along a narrow ledge until it crosses the neck of the boggy Alvnes promontory. Then, after crossing the outlet of Lake Valnesfjord, it passes through more open country with farms as far as Mjønes when again it occupies a narrow fjordside ledge. After Vågan, Route 80 moves away from the coast and cuts across the marshy Tverlandet peninsula. Here there is a junction with Route 813, a short gravel road leading to the famous Saltstraumen Eddy.

Saltstraumen Eddy is one of the world's strongest maelstroms. Four times a day, 80,000 million gallons of water foam and swirl through a strait only 500ft wide and 150ft deep which links the outer Saltfjord with the inner Skjerstadfjord. Seen at its best on the incoming current of a spring tide, the eddy is an awful sight. Ships caught in whirlpools have been smashed to pieces on the rocks while gulls circle overhead to dive on the thousands of fish which follow the bait-fish sucked into the sound.

Route 80 continues past Tverlandet along another narrow ledge backed by steep hills to Valosen, where the coast terrace broadens and is wooded or farmed as far as Bodø.

The Nordland capital has had a history of mixed fortunes. As early as 1803 some merchants from Trondheim set up a trading post on the peninsula which thrusts south-westward into the open sea, from which it is protected by the rugged islands Store and Lille Hjartøy. In 1816 the settlement, numbering a hundred or so, was granted the status of a town and took over the administrative rôle of the Bodøgård estate. It was hoped and planned that Bodø would become the major fish exporting centre for Nordland, but this proved over-optimistic. Only by chance, from 1864 until almost the end of the century, did the dream materialise when vast shoals of herring made the neighbouring seas their spawning grounds. The population grew ten-fold to 3,000 in what might be described as a fish-rush, only to stagnate again when the herring moved south.

Some advantages accrued to the town through the export of Sulitjelma ore, but the town's twentieth-century growth has been slow and attributable to its establishment as an administrative and commercial centre of some importance.

Comparable with Bodø's misfortune in losing the herring shoals was its destruction during the war. The scene of the Allies' withdrawal from Nordland (see Chapter 2), Bodø was almost completely destroyed in an air-raid of incendiary bombs on 27 May 1940. Its wooden buildings burned furiously, fanned by an on-shore wind. In re-building, concrete has replaced timber in most of the larger buildings, among which are the municipal offices occupying an imposing site near the harbour. The Lutheran cathedral, with its fine stained glass window, was completed in 1952 and dedicated in 1956, while the old church at Bodin to the south of the town dates from the twelfth century and has a seventeenth-century altar piece.

The town is a focus for communications. Apart from Route 80, the port is a stopping place for the Express Steamers. The railway which crosses the Polar Circle has terminated here since 1962 and ten years previously the airport was opened for all-weather flying. Its nodal position has given Bodø a large share in the tourist trade, as well as making the town an important military station in the NATO defensive network. The whole disc of the sun is visible at midnight during most of June and early July, the best view being from the restaurant situated on top of Rønvik hill to the north of the town centre.

Distances: Fauske to Saltstraumen Eddy: 36 miles/58km; to Bodø 40 miles/64km.

5. *Route 826 : Vargåsen to Røsvik*

This is an earlier section of the Arctic Highway which took the road out to the Røsvik ferry before the new stretch of Highway was opened to Sommerset in 1966. With the inevitable loss of traffic, Route 826 has been somewhate neglected and surfaces contrast markedly with the new Arctic Highway.

The road climbs away from the Fauskeidet depression up past the Kvitblik lake and reaches a height of over 250ft before descending steeply to Djupvik on the Sørfoldafjord. From Djupvik, the route uses the rugged shore of the fjord, which can accommodate only one or two farms on its steep slopes. Røsvik, at the end of the road, no longer has a ferry to Bonnåsjøen, but a link with Nordfold on the Steiga peninsula is maintained. By way of a short road (Route 81) of no more than twenty miles across the peninsula, then a ferry to Skutvik, this route re-joins the Arctic Highway at Ulsvåg (see also 6 below). Røsvik itself was founded as a trading post, Røsvik Hovedgård, as long ago as 1760.

Distances: Vargåsen to Røsvik: 15 miles/23km.

6. *Route 81 : Ulsvåg to Skutvik*

This is the continuation of the road described above. It leaves the Highway at the village of Ulsvåg and follows a largely shoreline path to Skutvik on Hamarøy. The surface of the road is water-bound gravel and, although there are no steep gradients, the route is not always well maintained.

The peninsular road is interesting. Numberless small fishing communities are passed and the scenery of high peaks on Hamarøy and the broken coastline is enchanting. The road also passes the childhood home of the novelist Knut Hamsun; a bust of the writer stands by the roadside.

Distances: Ulsvåg to Skutvik: 22 miles/36km.

7. *Route 814: Bognes to Korsnes*

This short section of gravel road was previously part of the Arctic Highway before the ferry point was transferred from Korsnes to Bognes. It leaves the Arctic Highway just outside Bognes and follows the steep shore of Tysfjorden for just over two miles to Korsnes.

Distances: Bognes to Korsnes: 2 miles/3km.

8. *Route 819 : Ballangen to Kjeldebotn*

Route 819 is a gravel road leading to the well-settled shores of Narvik's fjord: Ofotfjorden. The road keeps closely to the edge

of the fjord for most of the route and provides views across the very broad inlet. The Arctic Highway's junction with the road is at Ballangen village, the Highway taking the southern shore of the Ballangen arm and Route 819 following the north. At Kjeldebotn there are some minor, local tracks inland into Kjeldemarka.
Distances: Ballangen to Kjeldebotn: 13 miles/21km.

It should be noted that between Narvik and Tromsø there is an especially large number of branch roads off the 157 miles of Highway (see 9-22 below). Most of these routes lead west to the most heavily populated part of North Norway: the islands and peninsulas of south Troms. Others are in the Målselv valley, an area of early settlement and farming. In addition to the branch roads described below, there are very many minor roads and tracks of indifferent quality—a substantial number of these is closed in winter and during spring thaw.

9. *Route 19 : The Lofot Highway : Bjerkvik to Å*
Who could resist a road leading to a place called Å? But Route 19 is more than just a branch road from the Arctic Highway. It is a major route which, by ferries, bridges and secondary roads, is the artery of communications of the Lofotens, Vesterålen and Hinnøya. Its length and importance (it serves a community of perhaps some 100,000) are such that it cannot be done justice in these short pen sketches.
Suffice to say that in many respects it is as interesting as the Arctic Highway itself and, together with the whole of this unique arctic island realm, deserves a full description in a separate book.
Distances: Bjerkvik to Harstad (via the new bridge across Tjeldsundet and using Route 83): 53 miles/86km; to Svolvær (ferry to re-join Arctic Highway via Skutvik and Route 81); 149 miles/239km (plus 3 ferries); to Å: 223 miles/358km (plus 6 ferries).

10. *Route 84 : Fossbakken to Nordstraumen*
The road is one of many which leaves the Highway to serve

southern Troms. The first part of the route, as far as Salangen, was quite good even thirty years ago, but the rest is varied with some especially narrow sections. The whole road is water-bound gravel.

From the Fossbakken junction, the road follows the narrow and steep-sided Spansdalen to Tennevoll on the Lavangen inlet where Route 848 continues round the southern coast to the ferry point of Myrlandshaugen (16 miles/26km). The fjord shore is fairly broad and Route 84 passes a number of farms as it goes round the head to Lavangen before climbing over 600ft through the Lavangen depression. It then descends the Sagdal to a little inlet off Salangenfjorden where it is met by Route 851 (see 11 below).

The rest of the road to Nordstraumen is inland, first ascending through a marshy valley past Røyrbakvatnet, over a col and down the forested Bjørkebakkdalen. Here a short road (Route 852) branches off to Brøstadbotn two miles away on the coast. Route 84 strikes inland from the Elvevoll junction up towards Lake Skø, at nearly 600ft. After running along the north shore of the lake, the route follows the desolate valley of the Skø river which drains into Reisafjorden. The last part of the road, to Nordstraumen, is around the head of the well-farmed fjord.

This route is somewhat difficult during the spring thaw.

Distances: Fossbakken to Salangen: 19 miles/30km; to Nordstraumen: 52 miles/83km.

11. Route 851 : Brandvoll to Salangen

Leaving the Highway at Brandvoll, this road passes up through the exceptionally beautiful Salangsdalen towards lakes Øvre and Ner. It skirts the edges of these shallow lakes along a difficult path and crosses a short river which connects them. The whole road is very flat and usually well maintained. At Salangen it joins Route 84.

Distances: Brandvoll to Salangen: 12 miles/19km.

12. Route 87 : Elverum to Øvergård

This is certainly one of the most fascinating branch roads off the Arctic Highway. Only the junction with the Highway is asphalted, but the gravel surface of the rest of the route is adequate. Parts still have to be closed after heavy snow and again during the spring thaw. Although there are some very winding sections, the engineering is especially good and the gradients moderate.

Route 87 leaves the Highway at Elverum, crossing to the eastern bank of Barduelva and into the valley of the Målselv. The river itself is crossed at Rundhaug above its most contorted section where the meander belt is broad and the valley marshy. This part of the road is relatively recent, but at Rundhaug it is joined by the older Route 854 (see 15 below) and it is really a continuation of this road that Route 87 uses to Skjold.

The road keeps close to the right bank of Målselva, while mountains rise to about 4,000ft to the north and south. Small cirque glaciers burrow into the rock face on the north facing side. At Skjold, the road from Heia (see 17 below) joins Route 87, which itself continues eastward along the river valley towards the salmon-rich lake: Lille Rostavatn. The maze of local tracks and paths through the valleys in this area are frequently in a poor state. Near the western end of the lake, the road turns sharply northward to follow the river Tamok through one of the most captivatingly beautiful valleys in Troms: Tamokdalen.

Route 87 re-joins the Highway at Øvergård, making it a possible alternative to the Highway for the unhurried traveller as well as a useful local road.

Distances: Elverum to Skjold: 25 miles/41km; to Øverård: 48 miles/78km.

13. Route 86 : Andselv/Bardufoss to Finnsnes

Leaving the Highway to continue along the Måls river, this road turns north-westward to follow a steep, winding path up to Andselvatnet at over 500ft. This lake is enclosed by mountains and, to follow the northern shore, the road has been blasted out of the rock face. At the western end of the lake there are marshes;

the descent towards Nordstraumen on Reisafjorden is again winding but gentle along the Tømmer valley. By skirting the wooded eastern shore of another lake Reisvatnet, the hamlet of Nordstraumen, is reached and there is a junction here with Route 84 (see 10 above).

From the hamlet, the road follows the farmed shores of two arms of the Solbergfjord—Reisafjorden and Finnfjorden—to the village of Finnsnes. A ferry link with the large and sparsely inhabited island of Senja is possible across the Gisundet sound from Finnsnes.

Except for the last part of this route, near Finnsnes, which is asphalted, the surface and condition of the road are only moderate, but closure is rare.

Distances: Andselv to Nordstraumen: 15 miles/24km; to Finnsnes: 27 miles/44km.

14. Route 855 : Buktamo to Finnsnes

An alternative route to Finnsnes is possible by leaving the Highway just before reaching the Målselv bridge from a southern approach. The first part of the road is more recent than the section from Karlstad, where the old pre-war road is joined. The surface is gravel throughout and, although the condition is sometimes poor, there are no difficult gradients.

From Buktamo the road follows the river Måls along its left bank to Karlstad. Here a ferry still crosses the river to Gullhav and Route 854 (see 16 below), but its importance has greatly diminished in recent years.

At Karlstad, the road leaves the river and follows a marshy stream-filled gap to Finnfjord lake, which it passes along its northern shore to Finnfjordeidet. Here a secondary road (Route 856) follows the western side of Lake Rossfjorden to Rossfjordstraumen for about eight miles. In the next two and a half miles there is a descent of about eighty feet to Finnfjordbotn to join Route 86 into Finnsnes (see 13 above).

Distances: Buktamo to Finnsnes: 15 miles/24km.

15. Route 854 : Olsborg—south to Rundhaug

Instead of turning sharp left like the Highway at Målselv bridge, this road turns right to go up stream along the river terraces. Up to Nyberg, it keeps close to the river but then turns away and is separated from the Målselv and two falls (Bardufossen and Målselvfossen) by an 840ft peak. The falls feed an electric power station. The road re-joins the river at Flatmo, from which point it runs on a parallel path to Route 87 (see 12 above) along the meandering river to Rundhaug bridge.

This road, which is entirely oil gravel or asphalt, is usually kept in good condition.

Distances: Olsborg to Rundhaug: 14 miles/22km.

16. Route 854 : Olsborg—north to Navaren

This part of Route 854 branches off the Highway at Olsborg. When the Highway turns inland, this road continues downstream with the Målselv but at some distance from the river, almost to its mouth, to Navaren.

The route passes through well-farmed countryside, providing excellent views of the river and the steep western shore which includes the 4,000ft Vassbruntind.

Distances: Olsborg to Gulhav ferry: 8½ miles/14km; to Navaren: 16 miles/26km.

17. Route 857 : Heia to Skjold

At Heia a road branches south-east off the Highway to join Route 87 (see 12 above) at Skjold. Although in only moderate condition it passes through picturesque mountain landscapes and is subject only to short period closures.

The first part of this road is along the north shore of a large lake: Takvatnet. Near its eastern end, after a short climb past two other lakes to 800ft, the road crosses the lake's outlet and follows a winding path on a steep and difficult course down the narrow Langvassdalen to Skjold and Route 87.

Distances: Heia to Skjold: 13 miles/21km.

18. Route 858 : Storsteinnes to Oldervik

This short branch road takes the opposite direction to the Highway at Storsteinnes. Rounding the Sørkjosen bay on its northern side, it passes by a number of farms before reaching Skjæret where another road (route 859) continues along the fjord shore for another three miles.

At Skjæret the road turns inland through a broad forested gap, passing near to Josefvatnet before reaching an almost completely enclosed inlet, Stalvikbotn, littered with islands. The road terminates at Oldervik, by the bottleneck entrance to the inlet, but local tracks continue on a difficult path to the end of the peninsula.

Distances : Storsteinnes to Skjæret: 3 miles/5km; to Oldervik: 16 miles/26km.

19. Route E78 : Nordkjosbotn to Tromsø

This is one of the Arctic Highway's most important branch roads. It links the Highway with the fylke capital of Tromsø and (see 21 below) is part of an international road to Finland. It is not the traditional link between Tromsø and its hinterland (see Route 91 below), having been completed as late as 1936.

Only short sections of water-bound gravel remain on Route E78 and there are considerable lengths of asphalt near Tromsø and around Ramfjorden. Only in exceptional weather conditions is any part of the road closed. Its lack of gradient and some straight sections make it one of the fastest of the branch roads.

The route leaves Nordkjosbotn through Vollan and travels round the farmed shore of Balsfjorden, from which there are good views of high mountains to the south (see Chapter 5). At Kantornes it forsakes the fjord edge to drive a path northwards through two forested valleys separated by a broad col. This part of the route is known as Lavangsdalen but, in fact, comprises the valleys of the Smalatelv and the north-flowing Mellemdalselv. Near the 270-foot col between these streams is the Sara ritual stone, a reminder that this whole area has strong Lappish connections.

Page 175 Lapp boys wait to sell goods to tourists. These boys attended the
Lapp school in Kautokeino and spoke tolerably good English.

Page 176 A mobile hotel on the Highway outside the Bossekop post office.

Route E78 regains sea-level at the head of the right-angled Ramfjord, whose shore it follows, past the junction with the broad Breivikeidet and Route 91 (see 21 below) and on across a gap in the hilly Ramfjordnes promontory. Now it is back with the wide Balsfjord. Across the water, islands large and small block the view to the open sea and, on the landward side, mountains rise to over 2,000ft. Much of this fjord shore has been settled since land communications with Tromsø became possible. As the narrow sound between Tromsøya and the mainland is reached, one is already into Tromsdalen, a thriving suburb of Tromsø rather than a village in its own right (*see plate, page 121*).

Tromsø, another of the arctic's island towns, is more than just a fylke capital. It is the largest town in North Norway, with a charm and sophistication all its own. Its history of settlement stretches back over four thousand years and it has been an important trading and religious centre since the thirteenth century. The settlement received its charter as a town nearly two hundred years ago.

Tromsø's claim to be the capital of all North Norway, however, cannot be justified; communications are still such that an area as vast as Arctic Norway cannot hope to have an effective capital, and the influence of Tromsø is strictly limited. This is not to deny the town its leading position in the urban hierarchy of the North. Particularly since its absorption of a number of neighbouring kommunes in 1964, Tromsø has often been spoken of as the natural capital of the three northern fylker, as though this were an established fact instead of a pipe dream of the city administration.

In modern times the most important event in Tromsø's history has not been the attempt to set up an alternative national capital here in 1940, or the sinking of the *Tirpitz*, or even the development of the town as an academic and cultural centre. In terms of total and lasting effect, the opening of the giant bridge across the Tromsesund in March 1960 overshadows almost everything else in the town's long history of change. From the moment vehicles could cross its elegant 3,400-foot span instead of waiting

L

hours for the ferry it replaced, the expansion of the town and its surroundings has been explosive. Traffic across the sound doubled overnight and has continued to grow at a rate unparalleled in North Norway. There is a lesson here for the whole of arctic land communications.

Expansion has robbed Tromsø of some of its charm. Largely undamaged by the war, many of its nineteenth-century wooden buildings are now either eclipsed by characterless concrete blocks or in a state of disrepair. In May 1969 a fierce fire raged round the city square and along the main street, destroying property worth over 60m Nkr, including many timber buildings, and it will surely be only a short time before *old* Tromsø is no more. Some of the elegance and cosmopolitan atmosphere remain. Shops still sell polar bear skins and fine pewter, but the title *Paris of the North* seems less appropriate in the 1970s than it did twenty-five years ago.

Tromsø is served by the Coastal Express and it has its own airport two miles to the north-east at Langnes. It is a communication node and a centre for arctic studies, yet somehow detached from all but its immediate surroundings. Its claims to be the largest town in Norway—its administrative area is not far short of 1,000 square miles—but its size and near 40,000 population make it too self-contained and introspective to enable it to play its proper rôle in North Norway.

Thirty or more miles away, the Arctic Highway runs its course like a ship passing a headland light; the Highway experiencing the rigours of the arctic while the town stands secure in its island home.

Distances: Nordkjosbotn to Tromsø: 45 miles/73km.

20. *Route E78 : Road of the Four Winds : Oteren to the Finnish frontier*

As its name suggests, this road is part of an old Lapp route across to the Finnish panhandle. Improvements related to the new path of the Arctic Highway on the eastern side of Lyngenfjorden (see Chapter 8) and its oil-gravel surface in Skibotndalen support Route E78's claim to be an international road. Earlier widening of the road took place under the German occupation.

The *Road of the Four Winds* branches off the Highway at Oteren and follows the eastern shore of Lyngenfjorden to Bakken. Westward the snow-capped Lyngen Alps rise from the facing fjord shore and above the road tower the walls of mountains whose peaks exceed 4,500ft. At Storfjord, beyond the crossing of Lyngselva, three minor roads lead off into Signaldalen and Kittdalen (see Chapter 4).

The little village of Skibotn lies just over a mile from Bakken and, strictly, is off Route E78. It will, however, be on the new Highway. Skibotn is sited at the only major gap in an otherwise continuous mountain wall fringing the fjord. The break is caused by the Skibotnelv which drains a large glaciated valley, itself providing a route way into the mountains.

Today it is difficult to see Skibotn as anything more than just another arctic village. Yet it once had an important market dating from ancient times and its November and March fairs were still of considerable consequence when it was officially authorised as a market in 1840. Here Norwegians, Finns (Kvæns) and Lapps traded together. Here large numbers of Finns and Lapps crossed into Norway and settled around Lyngenfjorden when times were hard in the interior. Immigration was at a maximum in the mid-eighteenth century and Finnish surnames are common in the Lyngen district today. The big houses of merchants once contrasted with the simple huts of fishermen.

Lapps still use this route during their seasonal migrations and it was through here that the German army of occupation retreated into Finland in 1944.

From Bakken, Route E78 climbs through the wild and beautiful Skibotndalen towards the Finnish frontier. At points along the road the view back towards the village is breathtaking as the Lyngenfjord appears framed between the precipitous slopes of the valley. Before the war the road became no more than a track beyond Helligskogen, but now Route E78 presses on into the mountains. As the frontier is approached, the valley walls close in and the gradient steepens to bring the road to its highest point, 1,710ft, by the Galgovarre Lapp camp. Close to the boundary are some excellent examples of the strange *patterned ground* which results from permafrost. Perfectly formed stone polygons delight the eye of the geomorphologist or fascinate the puzzled casual traveller.

Beyond the frontier, the road becomes a major European route (E78 or Finnish Road 21) running along the Swedish-Finnish boundary. This road is of excellent quality. Not least among its attractions are farms where one can experience the *sauna* in its traditional form. At the twin villages of Karesuvento and Karesuando it is possible to cross into Sweden or, alternatively, one can rejoin the Arctic Highway via Enontekio and Route 93 (see 25 below).

Distances: Oteren to Skibotn; 19 miles/31km; to Finnish frontier: 41 miles/67km; to Stockholm: 1,020 miles/1,645km; to Helsinki: 815 miles/1,301 km.

21. Route 91 : Lyngseidet to Fagernes

This road is an alternative to Route E78 in that it links the Arctic Highway with Tromsø. It follows the more traditional path of travellers between the Lyngenfjord and Tromsøya and, for interest and scenery, is much to be preferred to the branch road from Vollan. However, there is little to recommend it if surface, width and general engineering are important criteria.

Leaving the Highway in the village of Lyngseidet, the road crosses the Lyngen peninsula by way of a broad valley to meet the long, narrow Kjosen inlet at the hamlet which shares its name. Across the fjord and towering above the road to the north are the impressive peaks of the Lyngen Alps. By the roadside the great Tyttebær gabbro screes provide excellent upper-surface gravels for roads in Troms. The north-facing mountains on the other side of Kjosen are etched by cirques whose glaciers reveal blue ice in the summer sun. This is one of those valleys which in winter presents a forbidding countenance but in summer, when the snow withdraws and the light returns, can only be described as majestic.

At Beinsnes the road reaches the more open Ullsfjord and travels along the shore to Svensby, where a minor road continues northwards. From Svensby a ferry (*see plate, page 122*) crosses the Ullsfjord to Breivikeidet. (It is also possible to cross the Kjosen inlet from here to Jøvik.) The service is not especially frequent, but a small café near the pier caters for hopeful passengers.

On the other side of Ullsfjorden the road first climbs gently up the broad birch-filled Breidvik depression. After crossing the river near Sandeggen, it avoids the marshy valley bottom by keeping to the steeply rising edge of Gabreilfjellet. At Fuglemo it crosses a steep ridge before descending to its junction with Route E78 (see 19 above) at Fagernes on Ramfjorden.

Distances : Lyngenseidet to Svensby : 15 miles/24km (Svensby-Breivikeidet ferry: approx 30min); Breivikeidet to Fagernes : 15 miles/24km.

22. *Route 864 : Olderdalen to Laukvoll*

This will soon be part of the Arctic Highway (see Chapter 8). At present, the road is being improved and simply serves the hamlets around Kåfjorden by following a largely shoreline path. At Kåfjordbotn a minor road leads off through the steep-sided Guolasjokka valley. The scenery along Route 864 is little different from that of the rest of Lyngenfjorden.

Distances: Olderdalen to Kåfjordbotn: 13 miles/21km; to Laukvoll: 22 miles/36km.

23. *Route 865 : Nordreisa to Bilto* (see also Chapter 5)

Another route destined to increase in importance in the near future is this road leading south from the Highway along Reisa-dalen (see Chapter 8). The path is close to the river, which is crossed about half way along the route at Hallen. Boats have used sections of the Reisaelv for centuries and the Lapps of the valley still forsake the road for the water during the summer.

The road is by no means well maintained but gradients are low and, in summer, the route is an attractive excursion from the Highway. Most of the journey is through forested terrain, but the lakes and waterfalls, Svartfoss and Puntafoss, provide diversions for the eye. Although the road stops at Sarelv outside Bilto, there are even greater attractions further up the valley (see Chapter 4).
Distances: Nordreisa to Bilto: 27 miles/43km.

24. *Route 882 : Langfjordbotn to Øksfjord*
This branch road off the Highway serves the fishing communities on the northern side of Langfjorden. While the Highway follows the southern shore of this fjord, Route 882 takes the northern bank as far as Tappeluft, then crosses a peninsular neck over a low col to join an inlet of Øksfjorden. Threading its way round the steep edge of the fjord, the road passes a number of small fishing settlements before reaching the Øksfjord village, close by the open sea at the fjord mouth. Across the fjord, glaciers crown the peninsular peaks. The Øksfjord glacier is the fifth largest in the country, while Svartfjelljøkulen, at 4,060ft, is the highest peak in Finnmark.
Distances: Lanfjordbotn to Øksfjord: 21 miles/33km.

25. *Route 93 : The Lapland Road : Alta to the Finnish frontier*
This road was completed as late as 1964. Before the war, the route, completed in 1938, terminated at Mieron and only snow roads led further south across the Finnmark plateau to Kautokeino and into Finland. Today, after considerable re-routeing, it is one of the three overland links between Norway and Finland, but, perhaps because of its later construction, it carries less traffic than the Karasjok road (see 28 below). Further, it provides an unequalled chance to see into the interior of the Finnmark vidda, into the heart of Lapland.

Some of the road is water-bound gravel with treacherously soft edges in the raised sections. The infrequent traffic allows vehicles to cut out a single pair of ruts in the surface; care is needed in departing from these tracks and driving on to unconsolidated gravel. The road dries rapidly in the prevailing conditions of low rainfall and strong winds.

A new section of road has been open for the last three years between Masi and a point three miles north of Mierojokka, and there are even more interesting developments taking place in the 1970s (see Chapter 8).

In the 115 miles between Alta and the Finnish frontier there is such a variety of physical and human landscape that it is difficult to select even the highlights. The road leaves Alta at the Bossekop *sentrum* and takes the general line of the left bank of Altaelva, although only occasionally is the river seen. At Skillemo, a minor road (part of the original route to Mieron) branches off to the east, crosses the Eiby river and provides an alternative route towards Kautokeino as far as Suolovuobme, where it re-joins Route 93.

This minor road is especially interesting, for it leads past the famous *Skifebrudd* slate quarries, owned by the State but worked by private enterprise, past the site of *Vina* (see Chapter 6) and keeps closer to the Altaelv than does Route 93. At first much of its path is along the giant river terraces; later it moves away from Altaelva, as the river sinks into Norway's greatest canyon carved 2,000ft into the plateau. The canyon was once a three-hour walk east from Bæskades but, today, a vehicle can be driven part of the way over a track leading to a wireless station. Both the surface and width of this minor road are exceptionally poor, the gradients are steeper than on Route 93 and the road is closed in winter—but, despite all this, to travel the length of the road is to see Lapland at its most untamed.

Route 93 follows the Eibyelv tributary of the Altaelv from Skillemo. At Kløftan is the rocky confluence with Vesterelva; the road then clings to the edge of a small gorge before coming into more open country along the gleaming ribbon of Lake Trangsdal.

The rest of the route is over the plateau, often well above 1,000ft, past dwarf birch and reindeer moss, bog land and glacial gravels—all cut by countless streams winding their way through a patchwork of lakes : wasteland except to the Lapps, for whom it is winter grazing for their reindeer herds, and except to the hundreds of Norwegians and Finns who fish the streams or gather the multer-berries.

After leaving Alta, the whole route is virtually uninhabited, except for an occasional Lapp hut or tent and for the two Lapp villages, Masi and Kautokeino. Masi lies just off the main road at the bottom of a hilly slope leading to the Alta river. Most of its buildings are post-war, following the vicious firing of the village one night during the German occupation. The total population is little more than a couple of hundred, but the village has its own clinic, church and school. Although its population is small, it serves a larger community when the transhumant Lapps return to the area in the winter. The house of the Hatta family is not only the telephone exchange but also the village gathering place for the coffee and gossip which is surely part of Lappish ritual. The little farms form a fertile oasis in the Finn-mark wilderness; there is not a more delightful village in all Nor-way. Yet, as I write, the village is threatened with submergence by a giant hydro-electric scheme on the Altaelv.

Kautokeino is much larger than Masi and probably older (*see plate, page 121*). Its present population is about 1,200, but this may grow as the nearby copper is exploited (see Chapter 8). The Lappish name for the village—*Guovdagaeino*—suggests its origin, for it means half-way-house : a resting place on the route between Lapp settlements on the Norwegian coast and those in interior Lapland, that is, in Sweden and Finland. From a winter encamp-ment, at least as early as the thirteenth century, it grew into a permanent settlement by the seventeenth century. Nonetheless, by 1880 there were still only a dozen or so houses, although the school catered for seventy pupils in winter. Today it is the most important Lapp settlement in Norway—almost all the popula-tion is Lappish—and the centre of the largest reindeer district.

The village shelters in a hollow carved out of the Finnmark plateau by the Kautokeinoelv (the Altaelv). The Route 93 bisects the village, leaving the church and school on the river side and the new tourist hotel on higher ground to the west. In winter the river is used for skating; in summer it provides a seaplane anchorage.

The first permanent building, a house for a priest and government officials, was constructed on a bluff on the right bank of the river in the mid-seventeenth century. In 1703 a church was built on the same bank on rising ground called *Goattedivva*—which means tent knoll—where a small hamlet of eight families had developed. Today's church, built in 1958, is close to this hillock. As with much of the rest of the village, the older church was destroyed by the army of occupation in 1944.

The school is the largest Lapp boarding school in Norway, or, for that matter, anywhere else. It enjoys, and deserves, a high reputation. Lapp parents as far away as Tromsø send their children to be educated here. Its swimming pool, library, cinema and ski jump provide for the needs of the local community as well as those of the pupils.

In 1968 a tourist hotel was opened just off the main road. It is certainly most comfortable and well-appointed, but one might doubt the wisdom of its siting (see also Chapter 8). Around the hotel, school and church are the scattered houses of the villagers. Few are now nomadic, but the reindeer industry remains important and much of the traditional way of life is retained. It is at Easter—when the village is full of Lapps from the surrounding winter pastures, when there are weddings at the little church and reindeer racing on the ice—that one appreciates that Kautokeino is the spiritual home of the Lapps, the Lappish Mecca.

There are two silversmiths in Kautokeino at the time of writing, but surely there is no stranger building in the whole of Lapland than the workshop of the silversmith and artist Frank Juhls. The house-cum-workshop is a short distance from the village on a track leading south west off Route 93 north of the Kautokeinoelv bridge. The outward appearance is of a large strangely-shaped wooden and glass tent. Inside the studio, which also serves as a living room, the music of Bach plays from a stereophonic record-player and all that is best in Scandinavian design elegantly furnishes the split-level rooms. Everywhere are the products of the small international colony of craftsmen who make up this curious community, at once alien to, and yet part of, Lapland.

South of Kautokeino, Route 93 presses on across the fjell to the frontier, past reindeer herds which scatter at the approach of trespassing vehicles. The most obvious sign that Finland has been reached is the sudden improvement in the road surface; like all north Finland roads, the continuation of Route 93 into Enontekio is of a high standard.

Distances: Alta to Suolovuobme: 32 miles/51km (33 miles by the minor road described above); to Masi: 43 miles/70km; to Kautokeino: 89 miles/44km; to Finnish frontier: 115 miles/ 186km; to Helsinki: 858 miles/1,381 km; to Stockholm: 1,074 miles/1,732km.

26. *Route 94 : The Midnight-Sun Road : Skaidi to Hammerfest*

At the Skaidi road junction, Route 94 branches westward, following the Repparfjordelv through its wooded valley to the mouth where the road bridges the river and runs along the barren southern shore of the fjord. After crossing a stream, the route continues northward to the Kvalsund headland, where a ferry fights its way across the fast-running Kvalsundet. This is an exciting five minute journey, the waters racing through the narrow strait which separates Kvaløya from the mainland. In spring and autumn, herds of reindeer cross the sound in their annual migration to the island. Landing at Stallogaro, the ferry passenger is reminded by the Lapps' sacrificial *Stallo* or troll stone that this is a reindeer summer pasture.

In fact the island presents such a bleak and uninviting prospect that, at first, it is difficult to believe there is any habitation. The road winds its way round the almost deserted and rocky west coast, suddenly to sight Hammerfest over the top of a col leading into the superb natural harbour which is the settlement's lure (*see plate, page 139*).

Take away the great shoals of fish which inhabit the Norwegian Sea—take away the Findus processing plant—and Hammerfest would surely die. Kvaløya means *whale island* and it was as a whaling station that Hammerfest was founded to rival Archangel, achieving town status in 1789. In 1807, a traveller to the harbour was able to report that 'whales were everywhere', but there were only nine houses and less than forty inhabitants. The dwellings of four merchants, a shoemaker's workshop, a customs house and a schoolhouse: this was Hammerfest. By 1875 there was still only one street of straggling houses, yet four years earlier, another traveller remarked on the hospitality afforded to strangers and found 'comforts and luxuries one would little expect'.

Hammerfest is now the home of about 6,000 people, whose economy is utterly dependent upon fishing. The comforts and luxuries today are in the well-stocked shops, the cinema, the excellent bookshops, modern churches and hotels. The town has an extraordinary pride. It is not just the claim, at 70°39′48″, to be the most northerly town in the world: almost everything here is *the most northerly,* from the Catholic Church to the Skogstua forest. It is not just pride in having been the first town in Europe to have its streets lit by electricity or pride in the presence of the meridian stone which commemorates the early nineteenth-century measurement of a meridian arc to the Black Sea. No, the real pride is derived from the immense sense of community that seems to characterise the people of Hammerfest. Few failed to return to rebuild the town after the war and their reward is their present prosperity.

The Findus deep-freeze plant dominates the town and employs nearly 1,500 people. The harbour—it is the best in Finnmark—is used by fishing fleets sailing under the flags of many nations, as well as by the Coastal Express. Overlooking the peninsula which encloses the inner harbour is Finnmark's leading hospital. Patients can be flown in by seaplane, although there are occasions—as I experienced after an accident some years ago— when the plane is out of action and a very rough ride, hundreds of miles overland, may be necessary.

Distances: Skaidi to Kvalsund: 16 miles/26km; to Hammerfest: 36 miles/58km (plus 1 ferry).

27. *Route 95 : The North Cape Road : Russenes to Nordkapp*

This is one of the most popular branch roads off the Highway in summer, when tourists make for North Cape, the most northerly point of mainland Europe. The route is in two distinct parts: from Russenes to Repvåg; and from Honningsvåg to Nordkapp. A ferry links the two sections.

The mainland road, to Repvåg, is of very recent construction and was only completed in the late 1960s. The standard of engineering is good, although some parts are narrow. The path is along the western shore of Porsangerfjorden and, because the road is close to the water's edge, except near Svartvik, there are wonderful views across this great inlet. A further extension is planned (see Chapter 8).

From Repvåg a ferry sails northwards round the Porsangnes peninsula to the island of Magerøya. The journey takes only about half the time of the alternative ferry from Russenes.

Sheltering in adjacent inlets on a narrow neck of the island are the twin settlements of Nordvågen and Honningsvåg. It is from the latter that Route 95 continues its journey to North Cape.

Honningsvåg (*see plate, page 140*), if it were not officially a village, would be the world's most northerly town! It is practically on 71° N, fully twenty latitude minutes nearer the Pole than is Hammerfest. Like Kvaløya, Magerøya is treeless and forbidding. Just as at Hammerfest, snow fences line the hillside above the port of Honningsvåg, protecting it from avalanches. The comparisons between these towns and islands are limitless. Both owe their livelihood to fish and the Findus plant at Hammerfest is paralleled by Honningsvåg's herring oil factories and the Fi-No-Tro freezing and filleting plant.

Route 95 takes a winding, hilly path across the reindeer pastures which are the fjells. There are strong Lappish influences here and, just north of Honningsvåg, Lapps encamp at Nordmannset in summer. The road climbs by steep gradients to over 1,000ft through some wonderful mountain scenery. Passing the beautiful Tufjorden, the road runs on to the North Cape.

Here come the tourists in summer to boast—with that famous Lapland traveller, Louis Philippe—a visit to the most northerly point in Europe, unaware of the validity of Knivskjellodden's and Kinnarodden's rival claims. In fact, North Cape's 1,000ft cliff of mica-schist is best seen from the sea, from which it rises dark and majestic. Above the Cape, on the plateau, one is less impressed by the commercialism of the *North Cape Hall*.

The Cape was named by the English voyager, Richard Chancellor, in 1553, in his re-discovery of the sea route to Russia which had been in use in Viking days. Another English association with the promontory was the adoption by Admiral Fraser of the title Lord Fraser of North Cape after the sinking of the *Scharnhorst* in 1942.

Distances: Russenes to Repvåg: 30 miles/49km (ferry to Honningsvåg: approx 1½hr); Honningsvåg to Nordkapp: 21 miles/34km.

28. *Route 96 : Lakselv to the Finnish frontier*

After Russia's annexation of the Finnish Petsamo corridor and before the extension of Route 93 (see 25 above), this road was the only link between Finnmark and Finland. At the beginning of World War II, a bridge was built over the river Karasjokka to replace the ferry, and Route 96 linked to the Finnish road system at Karigasniemi by way of the Anarjokka bru.

This route, although better surfaced and less subject to closure, is somewhat less interesting than the Kautokeino road to which it is an alternative. Despite its high altitudes at some points on the plateau, its path is often through scrub forest and views are restricted.

Route 96 leaves the Arctic Highway at Lakselv and strikes south through scattered hamlets which are part of the heavily settled head of Porsangen. Crossing latitude 70° N, the road climbs gently through birch forest to Nedrevatnet, a lake in the course of the Lakselv. It is this river whose valley is followed for much of the way to Karasjok. By Nedrevatnet is the Porsangmoen military camp, where tri-lingual signs warn motorists not to stop or stray from the road in the training area.

The road passes Nedrevatnet and Øvrevatnet near their eastern shores. To the west the high field rises by a seemingly vertical wall to over 3,000ft but, on this eastern side, the plateau reaches less than half that height. Near Øystad, ancient tracks lead off to Alta and to Valjok on the Tanaelv, leaving the road to ascend by increasing gradients through the narrowing valley. As the woodland thins, the immediate scenery becomes bleak marshland, but at the 1,000ft Rivnesvadda plateau the view extends to mountain peaks nearly 2,500ft in height to the west and east.

Descending again, the road crosses bogland to reach the tree-enclosed lake Nattvatnet, rising once more over a series of low ridges to its highest altitude, 1,150ft, just three miles south of the lake. The rest of the journey into Karasjok is along a steep, largely forested path, over streams and across marshes, a descent of 700ft in twelve miles.

Karasjok is a Lapp village and rival of Kautokeino. In many ways the two villages are alike. Similar in population, both act as trading and service centres for the reindeer herders. Karasjok claims the title *capital of Lapland,* but Kautokeino, somehow more essentially Lappish, is more deserving. Perhaps it is the even more straggling nature of the Karasjok village or its more overt capitulation to tourism, that rob it of the spiritual quality which is Kautokeino's attraction. Certainly, in the almost total destruction of the village in 1944 much of its pre-war charm was also lost.

A small museum and the old wooden church, completed in 1810, are Karasjok's main tourist attractions; to see the village as a Lapp centre it is necessary to be there in winter, in temperatures that can plunge to −25°C, when the nomads have returned.

The village will eventually become an important communications centre (see Chapter 8). Already Route 92 (see 29 below) gives a link with the Tana, and snowmobile routes run west and south.

Route 96 leaves Karasjok to follow the right-hand bank of the meandering Karasjokka which is, in fact, one of the head streams of the great Tanaelv. At the Finnish frontier it forsakes the Karasjokka for the Anarjokka, which it bridges to cross into Finland. As with Route 93, once into Finland the road improves and fast speeds are possible on routes towards the Gulf of Bothnia.

Distances: Lakselv to Karasjok: 46 miles/75km; to Finnish frontier: 57 miles/93km; to Helsinki: 866 miles/1,394km; to Stockholm: 1,143 miles/1,844km.

29. Route 92 : Tana bru to Karasjok

This route, at the time of writing, is incomplete (see Chapter 8); the central section between Lævvajok to Valjok is only partly built. The oldest part of the road is that nearest the junction with the Arctic Highway, while most of the southern extension from Karasjok was constructed in the early 1960s.

Route 92 leaves the Arctic Highway on the western side of the Tana bridge. Throughout its length it is never far from the great Tanealv and as far as Lævvajokka bru it closely hugs the left bank. West of Gæidno, the river forms the Norwegian-Finnish frontier and, as far as Roavvegiedde, a minor Finnish road runs parallel to Route 92 on the other side of the Tana. It is this minor road which joins the Norwegian road to Polmak, described in Chapter 5; a further link is made possible, this time with Route 92, by the ferry near the Finnish village of Utsjoki. After crossing the river here, the Finnish road leads due south to Inari.

Most of the road to Karasjok is through birch forest, but it is possible to appreciate the distinctiveness of the open Finnmark plateau to the north which contrasts starkly with the heavily forested and marshy lowlands of Finland to the south. At the confluence of the Tana and Karas rivers, the route turns westward along the Karasjokka to meet Route 96 just north of the Karasjok village.

The Tana, besides being Norway's third longest river, is navigable for most of its length, but there are some falls and strong rapids, notably at Storfossen, near Lævvajok, and the

Ailestryke, south of Valjok. The river freezes in winter and forms a winter road suitable for both car and sleigh. The salmon attract anglers to the river in summer and boat trips are arranged for tourists.

Distances: Tana bru to Roavvegiedde: 42 miles/68km; to Lævvajok: 62 miles/100km; to Valjok: 85 miles/137km; to Karasjok: 118 miles/189km.

30. Route 890 : Tana bru to Berlevåg

From the eastern side of the Tana bru, this route turns north-wards off the Highway to run downstream along the widening Tanaelv to the Leirpoll inlet off Tanafjorden. The valley is well settled and this eastern flank differs little from the left bank used by the Highway and described in Chapter 6. From Leirpollen, the road leaves the coast to cross the bare plateau which is the Varanger peninsula. Uninhabited except by Lapps in summer, the open tundra is a marshy lakeland across Kongsfjordfjellet. The mountain scenery is not unlike that along the Arctic Highway between Gandvik and Neiden; if anything, it is even more wonderful.

The arctic coast is reached at Kongsfjord and then followed westward through tiny fishing hamlets to the village of Berlevåg. Totally unprotected, this shoreline is pounded by giant waves which gnaw at the rocky coast. Small bays give limited shelter and at Berlevåg protective bars enclose the harbour. Twice the sea broke the Svartoks mole, but it was re-built in 1964 and, with the Revnes mole, a really large harbour gives refuge to the Coastal Express and the fishing fleet.

Close to the summit of the Kongsfjordfjell a minor road (Route 891) leads north-westward over the vidda to Båtsfjord, which claims to be Norway's largest fishing village. The proximity of the arctic fishing grounds, along with the natural harbours, allow such villages to survive in apparent isolation; others, notably Hamningberget and Syltefjord, have declined since the war.

Distances: Tana bru to Leirpollen: 28 miles/45km; to junction with Route 891 : 46 miles/74km (then a further 19 miles to Båtsfjord); to Berlevåg : 85 miles/137km.

31. *Route 98 : Varangerbotn to Vardø*

Apart from its interest, this road is attractive because of its excellent surface as far as Solnes and it is almost totally without gradient. It is one of the oldest route ways in Finnmark, through an area with strong Lappish affinities and a history of settlement as old as any in North Norway. The road beyond Vadsø has been partly re-built in recent years, but is closed for periods in winter. However, a snowmobile route exists parallel with the road from Skallelv to Vardø.

The route keeps very close to the fjord except at its most easterly end. Where the road runs a little inland, paths lead down to the small fishing settlements on the shore. As far as Vadsø, the coastland is often steep, with wooded or boggy slopes rising to nearly 1,000ft. The road runs, for the most part, along terraces which carry it at times to over 50ft above sea level. East of Vadsø, the cliffs are usually lower, standing back from the fjord to expose the wave-cut platform on which the road has its path. Here are all the features of an emergent coast, complete with former sea stacks rising high and dry as outliers of the cliffs. Elsewhere, wind blown deposits of sand have created dunes choking the sparse vegetation. For the geomorphologist the coast between Vadsø and Vardø is full of interest.

Most of the fishing settlements lie west of Vadsø and the main stretches of uninhabited shoreline lengthen eastward. Even some of the smallest hamlets have features of unique interest : Ekkerøy with its Viking graves, or Kiberg and Domen with their relics of the German occupation. Apart from the two main towns, however, one of the particular attractions of this route is a small Lapp museum close to the junction with the Arctic Highway.

This museum contains the private collection of Abraham Mikkelsen, knife-maker to the Finnmark Lapps. Consisting essentially of a reconstruction of the *amtsmannsgamme* which, for three hundred years to 1900, served as the administrative *building* for the district, the museum contains a comprehensive

M

collection of the Lapps' material culture. The turf hut, or gamme, was used by visiting officials as well as being a community meeting place, court house, chapel and even a school. Although the taxidermic exhibits might not be to everyone's taste, this is certainly one of the best rural museums in Lapland.

Vadsø is the fylke capital and adminisrtative centre. It has the disadvantage, by its easterly position, of being remote from most of the vast Finnmark territory, but it is a clean, proud town of about 3,500 population. The settlement was originally and typically on an island, Kirkeøya, off the shore but, in 1710, a church was built on the mainland and the village shifted its site. By 1833 it had been made a town by Royal Charter and it grew in size and importance as a fishing community. The war saw the town's destruction, but the re-building provided an opportunity to re-plan it and include some pleasant new buildings. Most impressive is the new church, on rising ground behind the middle of the town, and the large administrative offices. The town is well kept, with public gardens tended by school children and flowers grown under glass. The library houses a fine collection of ancient manuscripts of North Norway and North Finland. The latter are appropriate here where nearly half the population claims Finnish origin. The town's economy is, of course, based on fishing and fish products and the harbour, built around a natural promontory at the centre of the town, is sheltered by the island, now called Vadsøya.

At 31° E—the same meridian as Cairo—lies Vardø. A sixteenth-century English account of the settlement described it as a 'miserable place', a sailor even mistaking the tiny church for a reindeer! Today's church is somewhat more imposing and unmistakable, but Vardø certainly remains one of the most unattractive and depressing settlements I have visited anywhere. Unlike Vadsø, it has remained on its island site although, after the war, the population actually voted to re-settle on the mainland. The roads on the island, even in the town, are largely un-made, and the horse is as important as the motor car.

The town was originally built round a fortress—the Vardøhus

—and church in 1307 and grew to its present size through fishing and early trade with Russia. Today it has a population of about 3,000 and is connected to the mainland by a ferry across the Bussesund. Replacing the original fortress, an octagonal fort was built, on Vestøya, in the period 1734-8. Twenty fishermen-soldiers manned this: the only fortification on the coast of North Norway. After just over fifty years it was dismantled, but was re-equipped again in 1800 to protect the fishing fleets against Russian *interference*. This fort, together with the museum, are Vardø's sole attractions.

Distances: Varangerbotn to Vadsø: 30 miles/49km; to Vardø (Svartnes ferry point): 78 miles/125km.

32. *Route 885 : Hesseng to Nyrud*

Before the war this route was being developed as a major road to link with Finland's arctic highway. By the beginning of the war there were already three links with the Finnish road at Svanvik, at Skogfoss and at Grensefoss. A bridge was completed during the war at Grensefoss to replace the ferry. Today, this road stops at Nyrud, Russia has replaced Finland as Norway's neighbour and the Finnish links are gone.

The road leaves the Arctic Highway just before Kirkenes and follows a path along the general line of the Pasvik valley. Like the links in a chain, lakes are strewn across the path of Pasvikelva and the whole area is a marshy, undulating plateau. Some of the lakes, such as Bjørnevatn, have been drained, but the southern part of this route can be very soft in the spring thaw. Nearer the Arctic Highway roads lead east off Route 885 to the iron mining settlements, but the greater part of this road is through forest. The scenery can best be described as pretty, with quiet lakes encompassed by trees and carefully cultivated farmland. However, the route's real attraction is to the naturalist for much of the vegetation, with its alternating bog and conifer, can be compared with the Siberian taiga and the forest is a natural bird sanctuary.

Distances: Hesseng to Nyrud: 60 miles/90km.

33. Route 886 : Kirkenes to Grense-Jakobselv

Except for the first five miles, this road is narrow and of very poor quality. From Vintervollen, a snowmobile route replaces the road in winter. Before the war, the road stopped at Storbukta but was more important because it gave access to the Finnish arctic highway.

This is a route that leads almost nowhere, for from it run minor roads to the Russian border which is closed. Yet this is its bizarre attraction. Few visit Kirkenes without making an excursion—a pilgrimage—to the Soviet frontier. At some places it is possible to drive right up to the boundary which is marked by no more than a wire mesh gate across the road. In four languages one is warned, among other injunctions, not to fire shots or shout abuse across the border! At first sight it would seem easy to cross into Russia unnoticed by the simple expedient of walking around the gate. But perhaps that look-out projecting above the trees on the other side of the frontier really is manned.

In 1966 the border was opened to Boris Gleb but since then it has not been possible to enter Soviet Russia from Norway. By 1968, I found it impossible to obtain the permission of either the Norwegian or the Soviet authorities to cross into Russia here.

Distances: Kirkenes to Elvenes (end of main road): 5 miles/ 8km; to Grense-Jakobselv: 30 miles/48km.

Chapter Seven

The Highway and the Lapps

The total population of Norway north of Mo i Rana is some 400,000, of whom perhaps 2-6 per cent (depending upon definition) are Lapps. Some of the effects of the building of the Arctic Highway on the lives of the settled population have already been discussed, but the impact that this road has had on the nomadic Lapps is more subtle and difficult to assess.

Much has been written of the Lapps, of their origins and culture; much has still to be learnt. It is unfortunate that many accounts of these people have been written by those with a very limited acquaintance or without due regard to the unusual diversity among such a small ethnic group. Not only are there now clear distinctions between Norwegian, Finnish, Swedish and Russian Lapps, but within each country there are marked dissimilarities in groups from different districts, as between Karasjok and Kautokeino Lapps.

However, the most important contrast, as far as the Arctic Highway's influence is concerned, is that between the different economic groups. The Lapps are probably North Norway's oldest inhabitants and at some stage in their cultural development three distinct groups evolved. Initially, the Lapps were a wandering people: rootless hunters and trappers exploiting a hostile environment. Gradually, specialisations developed into more clearly defined economies as different locales proved more rewarding for one occupation than for another. Whaling and fishing attracted the attention of coastal communities, those inland found that domestication of animals eased the burden of daily life, but the

197

Lapps in the harsher expanses of the mountains and fjeld retained much of their old ways as hunting became pastoral nomadism.

It is not easy to explain the distinctiveness of these three Lapp groups. Especially difficult is it to understand why the mountain Lapps failed to become settled like their coastal and river brothers who clearly enjoyed a less arduous existence. No doubt some of the answer lies in the chance contacts made between Lapps and Scandinavian newcomers. In fact, many of the contacts can be traced through philological homogeneities, which can, in turn, be related to aspects of the culture which were shared by just such an encounter.

The Coastal Lapps and, to a lesser extent, the River Lapps are economically and, to a degree, socially integrated with the Norwegian population. The Mountain, or so-called nomadic, Lapps remain the least changed by time and contact. Theirs is an economy alien to the Norwegian; their settlements are geographically separate; they are the most Lappish of the Lapps. Not of them can it be said that in the twentieth century it is difficult to tell who is and who is not a Lapp.

Because of their isolation and distinctiveness, the nomadic Lapps have the greatest to gain and to lose by the development of communications. True, it gives new opportunities to expand and diversify their simple economy, but it also presents a threat to their heritage which, in the past, has been safeguarded by isolation. The more remote their communities have been, the less diluted has their Lapp culture become. It can be argued that, today, the *real* Lapps are confined to Finnmark (the land of the Finns, that is, the Lapps) and that, at best, the rest are Scandinavians of Lappish origin.

To fully understand how the Arctic Highway has influenced and will influence the nomadic Lapps, it is necessary briefly to describe their economy. Their whole life is intimately connected with their herds of reindeer, which often are their only source of wealth. The herds vary in size from about 300 to as many as 1,000 head. Remarkably, the subsistence level of the size of herds, at 300, has not changed much for centuries. Some of the herds, especially in Varanger, are swelled by deer owned by settled Lapps but tended by the nomads.

The deer find their winter pasture on the high fjells where mosses and lichens can be sought beneath the snow. As the sun returns to melt the snow in early May, the reindeer become restive and begin their instinctive migration towards the coast. Exactly why they move is not fully understood, but the complex reasons include the changing source of food and the millions of insects which inhabit the interior. In summer the lichens and mosses dry and would easily be crushed by the feet of the deer, while at the coast, grasses, fungi and the foliage of trees are available. Gadflies and midges are less of a plague in the more open and windy littoral areas than on the warmer bogland of the plateaux.

As the herds move, so too do the Lapps—out of their winter quarters, which are their homes, to temporary settlements on the summer pastures, which may be over 150 miles away to the north. Sometimes the deer will have to swim across the narrows which separate the mainland from the islands which will be their summer home. Here they will remain until the snow returns in late September or October, when the trek back will start. For centuries this has been the custom.

These migrations take place along traditional lines and the same summer pastures are used year after year. Indeed, it is correctly said that the Lapps should be called *transhumants,* for this is a primitive form of transhumance, not true nomadism. Nevertheless, nomad has such a romantic ring about it and so accurately describes the ancestral culture, that it is generally retained.

The organisation of the nomadic Lapps is based on the fylker which are sub-divided into districts with officials appointed on the recommendation of the herders. With a few exceptions, notably near the Russian border, each herd migrates from a winter pasture in one district to a summer pasture in another. The international boundary between Sweden and Norway is ignored and the boundary between Finnmark and Troms also is crossed twice yearly.

Exactly how many reindeer are involved is not clear, but reliable estimates suggest a figure around 100,000 (excluding Swedish deer using Norwegian summer pastures). Perhaps 1,500 Lapps are directly concerned. By far the largest numbers— 85-90 per cent of the total seems likely—are from Finnmark. Certainly, reindeer herding as the sole or main occupation becomes increasingly important the further one progresses north and east.

In Finnmark, the Arctic Highway either passes through the summer pasture districts or it cuts a path across the migration routes. It is in this fylke that the Highway has most affected the Mountain Lapps. Exactly what these effects have been is impossible to say—only time will tell—but they can broadly be divided into those which result from increased contact and easier movement within the community and those which arise from better communications with settled Lapps and the Norwegian population. It is impossible to generalise because some groups simply pass across the Highway twice a year and others, especially in Nordland and Troms, never reach the road. However, of those who cross the Highway, some camp near its branch roads and are as much influenced by road improvements as those whose pastures straddle the main artery.

The groups most affected come from each of Finnmark's four reindeer circuits (or groups of districts), namely: Kautokeino, Karasjok, Polmak and Varanger. The larger summer pastures astride the Arctic Highway include those of Sennaland, Børsfjellet, Ifjordfjellet and eastern Varanger. Also influenced by the Highway are the smaller districts (smaller, that is, in herd size) of Kvænangsfjellet (Troms) and of Saltfjellet (Nordland). Places where other road improvements have occurred in major pasture districts include greater Tromsø and Magerøya. Less affected have been the few small districts where migration is not practised.

Because communications have generally improved in recent decades, the migratory moves have been simplified. The practice used to be for the whole family to leave the winter quarters together and travel by sleigh and ski with the deer. This was a difficult journey for women and young children. The weather in May or in October is far from mild, there is snow on the ground and night temperatures plunge well below zero. The distance between the summer and winter pastures makes it necessary to set up intermediate camps along the route to obtain what little rest is possible, and tents must be hastily erected. The route hardly varies from year to year, often taking in traditional calving places from which the new-born deer will stagger within minutes of birth to follow the herd in its almost frenzied rush towards the coast. Few animals can match the speed of the reindeer and only its highly developed herding instinct allows some sort of control to be exercised by the Lapps accompanying them.

One is always struck by the hardiness of the average Lapp, man or woman, never more so than during migration. But now it is unnecessary for the whole family to expose themselves to the trek across the icy fjells. Instead, the men and some of the younger women go with the deer, while wives and children use public or private transport along the Highway and its branches. This has led to the decline of the traditional *pulka*, to the adoption of snow scooters and the Finnish sledge and the lessening use of the beautifully decorated treasure chests. The number of pack deer, too, can be reduced now that much of the family's goods and chattels can be transported between winter and summer quarters by road. On the move to the coast, the men and deer go first, leaving the women and children to follow; in autumn the order is usually reversed so the winter, or permanent, home can be prepared for the return of the menfolk.

Although this change in practice has obvious advantages, something has been lost. No longer is migration a family affair. No longer do the women tend the calving does. No longer is it the rule to change from winter to summer equipment along the route. After the Easter festivals, the family splits up, only to be reunited, perhaps two weeks later, in the summer quarters. Of course, not all families have abandoned the old ways, but the coming of roads has seen the demise of much that was traditional.

The vast improvement in communications between summer and winter pastures permits much more in the way of household utensils, clothing and so on to be carried from one to the other. Indeed, in the last few years there has been the occasional return to the interior during the summer by some Lapps in order to collect goods left behind or simply to do work on their winter quarters. Previously this was unknown; the move was final and there was no contact between the Lapp villages on the plateau and the temporary camps near the coast. Those settled Lapps of Kautokeino or Karasjok never saw the summer homes of their nomadic cousins. Only a few years ago, I took a Lapp woman from a vidde village to see relatives encamped by the Highway. Typically, she had never made such a visit before or seen the summer camps of those who, in winter, were her neighbours.

Partly because of the easier transport afforded by the Highway, the nature of the summer quarters has changed. The tent is only very rarely used, except as a store place or as a *tourist lure*. The turf gamme survives in only one or two districts, such as on the Kvænangsfjell (*see plate, page 157*). Instead, the Lapps have constructed wooden cabins in the summer grazing lands, to which they return annually. Most are simple two-roomed huts, at best indistinguishable from the summer chalets of Norwegians, but often small and tumbledown—showing the ravages of the winter climate. Materials for construction, repair and maintenance can be brought along the Highway, and it will surely not be long before these cabins have all the colour and brightness of the wooden houses in the villages.

The summer pasture lands were established long before the roads. The Highway and its branches often bisect the areas used by the herds of a group of families. Now that the Lapps attempt to keep herds separate from one another by wire fences, a special problem has arisen. Unless the groups of herds are pastured only on one side of the road, it is impossible to contain them. Nothing will prevent the deer moving from one pasture to another by getting on to the road and passing through the break in the fence lines made by the Highway. Various reindeer scares have been tried, the usual practice being to leave an item of clothing attached to the roadside fence in an effort to frighten deer away from the gap. One wonders if the day will come when grids are incorporated into the Highway's surface!

One of the most important social consequences of improved roads has been their polarising effect on the summer camps. Today, some of the Lapps have trucks or can use buses such as the North Norway or North Cape services, causing the roads to have a markedly magnetic effect on the siting of the cabins. Previously tents had been set up in a widely dispersed fashion on the tundra moorland. Water supply and a dry site were the main criteria when the Lapp family built its gamme or erected the tents. Now it is important, almost necessary, to be close to a road; if close to a road, why spread along the road, why not a summer village?

Just such a change has occurred along the Highway at Senna-land. Something akin to a village, with its own slaughter house and corral and its own church, has grown up at Aisaroiui (see Map 6). Nearly twenty families live here alongside the Reppar-fjodelv, allowing for contact and a community life unknown in the past. The same sort of agglomeration is affecting the summer camps on Ifjordfjellet and the Børsfjell. On both plateaux, the chapel, built near the Highway, is acting as a focal point, and on Ifjordfjellet much care has been given to the construction of one of the most delightful of all Lapp churches in Finnmark. This wooden chapel resembles two ridge tents built end to end and, set on a frame some little way away, a bell to summon the faithful.

The Highway allows more than just the families living near the chapel to attend the infrequent services. At Aisaroiui, for example, a Lutheran minister visits the village on four Sundays during the summer. The visit is quite an occasion. Everyone in the district will put on his best clothes and jewellery and make the effort to get to the service (*see plate, page 158*). Marriages tradi-tionally take place at Easter in the winter village, but a Christen-ing may well occur in summer. With the service being given consecutively in Norwegian and Lappish, the ceremonies will last much of the morning. Then follows coffee and gossip in the cabins of friends and relations. Those living some distance from the chapel may well have to thumb a lift back home in a passing car.

The sense and reality of community life in a quasi-village such as that at Aisaroiui is the Highway's greatest contribution to the nomadic Lapps. The road has given these people the chance to live together during the long days, the chance to develop and maintain the spirit of group co-operation which is signified by the Lapp word *siida*.

Additional material advantages may be gained by the en-camped Lapps in the future if the compactness of the new settlements help to persuade the State to provide the cabins with electricity from the power lines which cross the tundra.

On the debit side, the Highway has brought real danger. Al-though traffic is never heavy, accidents are not unknown. The deer can usually look after themselves; in fact they are more likely to cause a motorist to drive off the road than cause a collision. Children, however, do stray on to the Highway, particularly when playing near tents set up at the roadside. On a single day in 1967, while I was living with some Lapp families, no less than two children were involved in road accidents on a short stretch of the Highway east of Alta. One child was killed and a small boy was injured. If an accident occurs there is considerable difficulty in summoning help.

The new road system may well encourage the breakdown of the old districts. Already areas with better summer pastures are attracting the attention of groups whose traditional areas are either over-grazed or particularly susceptible to severe weather conditions. Some of the best districts are to be found in West Finnmark and on the Varanger peninsula, while vegetation in districts used by some of the Karasjok Lapps is especially sparse. Similarly, there are significant variations in the summer districts of Troms and Nordland. Troms lacks the more open vidder that characterises parts of north Nordland, but in neither of these two fylker is the pressure on grazing land as acute as in Finnmark. In fact, if it were not for the migration of Swedish deer into Nord-land in summer, some of the good pastures north of Mo i Rana would be little used.

The advent of the Highway and associated roads has had little effect on the nomadic Lapps in winter. Although the reduction in road closure periods is bound to lessen the isolation of the Lapp villages of the interior Finnmark vidda, winter has never been a season when roads really mattered. Skis, sledges and snowmobiles and the winter roads across the plateau reduce the importance of conventional fixed-line highways. Only with the construction of east-west plateau roads (discussed in Chapter 8) is it likely that winter life will change.

Just as the intra-community effects of the Highway, described above, are restricted to the summer months, so too, the increased contacts between the nomad or Mountain Lapps and the rest of North Norway's population and visitors occur mainly in the sum-mer season.

One of the most conspicuous effects of the Highway is the Lapps' involvement in the tourist industry. There are few very clear signs south of Lyngenfjorden but, just north of Djupvik, Lapps from Tromø sell goods in the tourist season. Here, against the dramatic backcloth of the snow-capped Lyngen Alps, is probably the most undisguised attempt of all to benefit from the increasing number of visitors to North Norway. The site is at a point on the road where cars and even buses can pull off the Highway. A tent or two, some simple wooden racks: this is all that is needed and all there is. But this is not a genuine Lapp camp. Some of the Lapps are settled in the Tromsø district and their cars are half hidden behind the snow fences close by. Prices here are higher than elsewhere and the Lapps bargain un-ashamedly with their customers, showing none of the reticence which so characterises their race. Only near Kautokeino and at Karasjok and on Magerøya is there such open selling.

Along the rest of the Highway the sale of goods shows little in the way of commercial enterprise. True, most of those in nearby summer camps will pitch a tent by the roadside and display skins to attract the tourist's attention. But, if it rains, the tent—and the reindeer skins—will probably be abandoned for the shelter of the cabin. Children frequently act as salesmen (*see plate, page 175*). They are less shy than their parents, are more willing to speak Norwegian and strike just as hard a bargain.

The most frequently offered goods are undressed skins, articles made from reindeer skin and deer horn. Quality varies, but most of the good autumn skins are sold to dealers and appear in the shops of Oslo and Trondheim. The tourist will more likely buy a skin which, when unpacked, will leave him with a strong piece of leather and enough loose hair to fill a pillow. A big set of antlers sells for a high price, but all manner of pieces of horn are cleaned and sold. The women make purses, decorated with ribbons, and slippers out of reindeer skin during the idle summer days (*see plate, page 158*). The cost of goods depends on the relative bargaining powers of the Lapps and the tourists. A large skin may cost up to 60Nkr but smaller ones sell for as little as 30Nkr. A full set of antlers, which will probably be tied to the roof of the tourist's car to advertise his arctic excursion, is expensive but more durable than a skin. If less than a full set will satisfy, then, at no cost at all, the tourist could obtain what he wanted from the litter around the slaughter houses set away from the Highway.

Children delight in their contact with motorists on the Highway, especially if they may practise their excellent school-English. Not to be out-done by their elders, they will produce items of their own for sale. Even crude drawings of reindeer and sea shells are thought worthy to offer the visitors.

This contact is obviously financially rewarding to the Lapps camped along the road. The receipts from sales of skins and trinkets may not be very large, but they are a valuable supplement to the family's total income. However, there is something unreal, tangential, about these meetings of Lapp and traveller. Many of the tourists do not speak Norwegian let alone Lappish. No relationship is established beyond that required by the business of buying and selling. The tourist will not see into the smoke-filled tent with its reindeer meat strung across the roof like drying laundry. The turf hut or gamme of the Kvænangsfjell remains a mystery, unexplained and unexplored. Little of Lapp culture brushes off on to the tourist bent on reaching Tromsø or North Cape with no time to spare. Even though their costume excites curiosity and is faithfully recorded on countless rolls of film, it is doubtful if the average tourist appreciates anything of the subtle regional differences or the functional nature of much of the Lapp dress.

To know just a little of the Lapps one must live with them; yet this is a privilege denied to all but a fortunate few visitors to arctic Norway.

The Highway gives the nomadic Lapps access to towns and villages in summer. Especially popular are Lyngenseidet, Alta, Lakselv and Varangerbotn. Transport is readily available to those owners of large herds who have a truck—probably a *VW*—and to their friends and relations. Others use the North Norway Bus or travel with the so-called *Reindeer Policeman* who controls the sale of meat and generally looks after the Lapps' interests. A few are bold enough to hitchhike, sitting silently in the back of the car, unaware that the foreign driver may well be unnecessarily apprehensive of the long knife which is every male Lapp's third hand.

The excursions to town or village are infrequent and may be made for a variety of purposes: to shop, to drink mild beer in the cafeteria of the local hotel or to call on relatives who, recently or in the remote past, retired from the rigours of reindeer herding. Once in the town or village, the nomadic Lapp keeps to himself or converses only with his own kind. A particular café or shop is often the meeting place of nomad and settled Lapp.

One of the most popular settlements for such visits is Alta, with its long association with things Lappish. Any day in summer will see a knot of reindeer Lapps just standing at the Bossekop crossroads or engaged in conversation in the entrance to the Alta *Samvirkelag* supermarket. Some wander down to the airport, but only a few will ever travel outside Finnmark.

During the months in the summer pastures, there is little work for the men. The reindeer look after themselves most of the time and there is opportunity, therefore, for the Lapps to fish or collect berries. These can be sold in the village and the cash used to buy provisions or alcohol. The Lapp is not an exceptionally heavy drinker but, as so often, with occasions restricted, opportunities are eagerly seized upon. Relaxed, his inhibitions washed away, the Lapp man may be heard *yoiking* a strange melodic tale—half song, half mouth-music—which is a part of his cultural heritage rapidly being lost. Because of its association with drunkenness, *yoiking* is frowned upon by many Lapps and my own recordings of children's *yoikes* were made only with much bribery and in great secrecy! The haunting sound of the *yoike* has much in common with the songs I have heard accompanying the drum dance of the Eskimo in east Greenland.

Many Lapps go to Alta to order belts and pouches from John Mienna, the son of a former nomad, who designs and makes these essential items of the male Lapp's costume. Others travel to Varangerbotn to buy knives from Abraham Mikkelsen, an excellent craftsman. It must not be thought, however, that the Lapps are constantly on the move during the summer. These excursions to the villages are rare, red-letter days. Few children ever get far from the cabins. They watch the twentieth century pass by their doors along the Highway and remain remarkably immune to its influence.

No traveller on the Arctic Highway in summer can help but be reminded that Finnmark is Lapland. Even in Troms and north Nordland, the Lapp, because of his costume, can readily be distinguished. To the nomadic Lapp the Highway has indisputably brought great benefits, easing the burden of his life in the summer reindeer pastures. All the same, there remains the uneasy feeling that the contacts it has allowed must inevitably erode into the age-long traditions of the Lapps. Many of the old crafts are dying. When the village shops can supply almost every need, there is little encouragement to carve coffee cups and butter dishes or even to continue to make all the family's clothing. How long will it be before the Lapp costume is abandoned, before reindeer herding ceases to involve transhumance, before the nomadic Lapp is finally and fully assimilated into the Norwegian population? Only time will tell. Isolation is the great safeguard of the traditional way of life and it is isolation that the Arctic Highway is committed to defeat.

N

Chapter Eight

The Future
with a note on tourism

It is not only in the use of terms such as *branch road* or *trunk road* that the dendritic nature of a highway system is recognised. Roads are like living organisms: they require careful cultivation and pruning; branches can wither and die; the process of evolution is never done. So, too, is it with the Arctic Highway.

The Highway is a young road and the changes of the 1960s have been as important and far reaching as the very construction of the road thirty years earlier. The rôle of the Highway in the development of North Norway in the coming decades is likely to increase in importance as improvements are made and the complementary system of roads linked to the Highway becomes at once more dense and of a higher standard. These developments clearly depend upon the availability of capital, upon social and economic forces such as mineral exploitation and tourism, and upon defence considerations. Reference to some of the economic problems and consequences has already been made in Chapter 1. Plans in the drawing offices today may be in the waste bin tomorrow. This chapter will simply look at the probable future in store for the Highway in the 1970s and 80s and only hint at some of the more distant prospects.

It is now quite clear that road communications in North Norway are slowly assuming an importance which, in the first half of this century, would have seemed unlikely and unnecessary. This region can no longer turn its back on the interior, to rely on the sea as a highway and to concentrate the greater part of its attention on a littoral economy. To do so now would be eventually to drain North Norway of its population in an exodus to the south and to create a barren waste out of potential richness. It would be ungratefully to undo the centuries of pioneer development for which generations of Norwegians, Lapps and Finns have been responsible. To adapt to the twentieth and, still more, the twenty-first century, North Norway must diversify its economy and comprehensively exploit its resources. In doing so, communications will play an all-important rôle. Sea traffic will continue to be vital with the Express Route, now nearly eighty years old, usefully serving the islands and fjords. Airways will expand at an even faster rate than in the 1950s and 60s. But, with the development of railways seemingly halted, it is the road system that will prove to be the infra-structural element of greatest importance. In this, the Arctic Highway, as the main artery, will dominate. (The only new rail route that may prove viable is a northward extension of the Finnish lines which serve the mining settlements at Kolari and Raajärvi. If Finland and Finnmark were ever to be linked by railway, then it would probably be the result of mineral exploitation on a vast scale in Finnish Lapland and the need to export ores from an ice-free coast. This would transform Alta or Lakselv, for instance, into another Narvik. Such an occurrence is highly improbable but certainly not impossible.)

As recently as 1968, the Arctic Highway was designated simply as Route 6 (previously Road No 50). It was thought inappropriate to use a European classification beyond Hell, north of Trondheim. Now, just a few years and many improvements later, R6 has become E6 as far as Lyngenfjorden. The Arctic Highway, or at least part of it, is a European highway.

Just what makes a European road may not always be very clear, but truly to achieve this status for its entire length from Mo i Rana to Kirkenes, the Highway must complete three tasks:

    i. The raising of the constructional standard of the road.

ii. The opening of the Highway to winter traffic on all its
sections.

iii. The elimination of ferries.

The first task is a continuous process limited by time and
money. The present aim of the Highway authorities is to asphalt
all the most heavily used sections and replace water-bound
gravel with oil-bound gravel. At the same time, sections are to
be straightened, widened and, where necessary, raised to reduce
accumulations of snow. The general scheme, as far as the Highway is concerned, is to proceed with improvements from the
south towards the north except where local conditions make immediate repairs or alterations necessary.

Anyone who has travelled on the Highway for a number of
years ceases to be amazed at the speed at which surfaces are improved. Today's broad oil-bound gravel road was yesterday's
pot-holed track. Recent improvements west and south of Olderfjord on Porsangen and in the Målselv district are typical, as
were earlier surface changes over the Saltfjell. There are, however, still many hundreds of miles of water-bound gravel requiring attention which, as explained in Chapter 1, is necessarily
protracted. When capital is made available there can be rapid
advances, as demonstrated by the speed at which much of Route
93 was recently re-surfaced.

The figures in Table I below give some idea of the generally
slow rate of progress in re-surfacing. They refer to all State
roads; the regions have to wait their turns according to priorities
which are subject to annual change.

Naturally, North Norway and the Arctic Highway have only
a small share of these improvements. Annual surfacing programmes are usually of the order of only 100-150km of asphalt
and 200km of oil gravel. The likelihood of the Highway having
oil gravel or asphalt surfaces throughout its length before the
mid-1980s seems remote.

As well as surface improvements, the Highway will be widened. The narrowest sections, chiefly in Finnmark, can hardly
accommodate a single large vehicle and, with soft shoulders,

	1968	1970
Total length of roads	23,665km	24,118km
	(approx 14,600 miles)	(approx 14,990 miles)
Hard surface (asphalt,	5,673km	6,370km
concrete, etc)	(approx 3,525 miles)	(approx 3,960 miles)
Oil-bound gravel surface	4,250km	6,257km
	(approx 2,640 miles)	(approx 3,890 miles)
Water-bound gravel	13,742km	11,491km
surface	(approx 8,540 miles)	(approx 7,140 miles)

Table I: State roads and their surfaces: 1968 and 1970

accidents are inevitable. The minimum standard width for the future is 7m (approx 23ft), with a standard of 8.5m (28ft) demanded of sections where projected densities in the 1990s reach 5,000 vehicles per day and 10m (33ft) where projections indicate over 5,000vpd. At present, much of the Highway falls far short of these requirements. As traffic densities increase, the Highway must be widened. Raising or re-surfacing a section will have to be accompanied by increasing the width. In the past, as for example over Saltfjellet, this has not always been done, but elsewhere, as east of Skaidi, the future has been given greater consideration. Some of the fjord sections will prove to be the most difficult to widen because of physical restrictions imposed by the narrowness of the coastal ledges, while on the trans-fjell passages the problem will be increased widening costs owing to the raising of the road above the plateau surface. Fortunately, the longest single section requiring widening, between Varangerbotn and Kirkenes, is likely to retain its low densities for some time to come.

Re-routeing of the Highway on any large scale is improbable for two reasons. Obviously, the cost of any radical change in path would be prohibitive and the original choice of route has proved to be a satisfactory compromise. Even if the coast route, to be referred to later, is constructed, the Arctic Highway will continue to be the only road which can adequately serve both coastal and inland communities.

Local alterations in the Highway's route, however, are quite likely. These will continue the programme of straightening and curvature reduction which has been followed for many years, but may also include some major changes in the more congested sections. A possible change in the route through Mo i Rana, for example, has already been mentioned in Chapter 3. A similar re-routeing in Alta, to take account of the town's planned growth, has been the subject of discussion in recent years. The heavily-used sections of Highway in Sør-Troms, especially between Brandvoll and Olsborg, may well see changes in the 1980s.

In Finnmark, one may speculate upon the feasibility and desirability of inter-fjordbotn routes to excise the long peninsular paths. Two such routes might be considered: between Kvænangsfjorden and Kåfjord, and between Alta and Lakselv. Today's Highway is about 115km, or over 70 miles, between Kjækan on Kvænangsfjorden and Kåfjord near Alta. A track leads east from Kjækan south of the Didnovarre range and into Mattisdalen, where a minor road joins the Highway south of Kåfjord. Although this track rises to over 1,500ft and winter conditions are severe, the distance between Kjækan and Kåfjord would be halved by its adoption by the Highway.

Even greater savings in distance would be effected by a more direct route between Alta and Lakselv. These two small towns, of growing importance, are scarcely 60km, or about 38 miles, apart, as the crow flies. By the Arctic Highway the distance is nearly three times as great. A possible linking route might follow the Transfordal away from Rafsbotn, cross the high divide of Ravtasjavrečokka and then use Stabbursdalen to descend to Porsangerfjorden. Alternatively, a track linking Tverredalen, across Adnevarre, with Route 96 south of Lakselv, already exists.

All the same, apart from the cost and physical handicaps of such new paths as these, there would be strong arguments in favour of the Highway continuing to serve the more populous coastal areas, leaving inland communications to a largely radial system.

The combined effects of re-surfacing, widening and straightening will be twofold. The capacity of the Highway will be increased to meet the demands of higher traffic densities: asphalt surfaces are expected to support axle loads of not less than 10 tons and oil-bound gravel minimums are 8 tons. Secondly, average traffic speeds will be raised by perhaps some 10 to 15 mph, or 15 to 25 kph, in good weather. Because the real growth sector in traffic using the Highway in the future will be long-haul, even small increases in speed will bring significant decreases in journey times.

The second task for the Highway authorities is to reduce winter closures to an absolute minimum. No arctic road can be expected to allow the passage of vehicles every day of the year. Sudden and heavy falls of snow take time to clear and some weakening of the surface and shoulders of roads is inevitable in spring. A large permanent maintenance staff has to be augmented by casual labour in time of difficulty. The authorities' primary concern must be to eliminate the period closures which were looked upon as unavoidable when the Highway was built. Ten years ago, five lengthy sections were closed in winter for considerable periods. Gradually these have been reduced to two as first Kvænangsfjellet, then the Polar Circle crossing and finally Sennalandet were kept open through the winter snows. Although annual and geographical variations in closure must be accepted as dependent upon the vagaries of the climate, future attention will be given to the two fjells, Børselv and Ifjord, which still have to close in the winter season.

The Highway on both Børselvfjellet and Ifjordfjellet will need considerable improvement before winter opening is possible. Both plateaux sections are high—over 600 and 1,000ft respectively—and the climate is transitional between arctic and subarctic. The Highway, surfaced with water-bound gravel, is narrow and largely unraised. Only with constructional improvements can it be hoped to maintain a flow of traffic in winter. No forecast can be made as to when winter opening may be possible. Much depends upon the general rate of progress of reconstruction programmes and the demand, which, it must be admitted, is not presently very convincing in either of these two areas.

Apart from local and transitory weather problems, the Highway is certain to have to grapple with the difficulties of its arctic location on all of the five plateau sections mentioned above, as well as along Varangerfjorden, every winter in the foreseeable future. Nevertheless, the success with which this fight has been waged in the past is encouraging for the coming decades.

When the Arctic Highway was completed in 1940, it relied on ten ferry links and two further ferries as alternatives to the road between Mo and Kirkenes. Today only four ferries are necessary and, before 1980, only one will remain (see Table II).

The next ferry to become redundant will be at Skjomen, south of Narvik, between Skjervik and Grindjord. The bridge, described in Chapter 4, is due to open to traffic in 1972. Although this is a very short fjord crossing, it is on a stretch of the Highway which carries a quite high level of traffic, so its elimination will be especially welcome.

A more ambitious project is due to be completed in the mid-1970s. For centuries, land routes around Lyngenfjorden in central Troms have ignored the steep eastern shore north of Skibotn and south of the Kåfjord inlet. The tiny hamlets that have found a place on this inhospitable part of the coast have been served from the sea. The Arctic Highway originally crossed the fjord from Årøybukt to Nordmannvik. Soon after the German occupation in 1941, both ferry points were abandoned for alternatives further south, and the more logical Lyngseidet-Olderdalen link was forged. Now a new road round the eastern side of the fjord is in an advanced state of construction.

The southern part of this new section is already built, for the Highway will use the improved E78 from Oteren to Skibotn (see Chapter 6). Thus the whole of the road from Nordkjosbotn to Skibotn will be re-designated E6/78 and the Arctic Highway will have a very direct link with Finland through Skibotndalen. From Skibotn an entirely new road is necessary to take the Highway to Laukvoll on the Kåfjord inlet, where it will join Route 864 which has been improved. This road will then be re-numbered as part of the Arctic Highway which it presently joins at Olderdalen (see

Chapter 6). The only road west of Laukvoll has been a minor *fylkesvei* serving Kjerring and Manndalen. The new road has

In 1940	By 1960	By 1970	In the future
Rognan-Langset	Replaced by road around fjord-head	*	*
Røsvik-Bonnåsjøen	No change	Reduced by new road: ferry—Sommerset-Bonnåsjøen	To be replaced by new road or bridge : 1980s
Korsnes-Skarberget	Ferry re-routed : Bognes-Skarberget	No change	No definite plans: ferry likely to remain
Sætran-Forså	No change	Bridges and new road	*
Skjervik-Grindjord	No change	No change	Bridge : 1972
Ankenes-Narvik	Bridge	*	*
Vassvik-Øyjord	No change	Rombak bridge	*
Årøybukt-Nordmannvik	Ferry re-routed : Lyngseidet-Olderdalen	No change	New road : 1974-7
Sørstraumen-Badderen	Replaced by road around fjord-head	No change	Bridge : 1974-7
Skallenes-Lakselv	Bridge	*	*

Table II: Ferries and their replacements on the Arctic Highway
(for further details see Chapters 2 and 8)

encountered difficulties in the relief of the eastern shore. Even the narrow ledge which separates high mountains from the fjord on the western side is missing here. Much of the road has, therefore, to be blasted out of the sheer mountain wall which plunges directly into the fjord. Tunnels, including a 550m length opposite Sandvika, have proved necessary but have generally been kept short to reduce costs. Avalanches are as common on this shore as between Elvevollen and Kvalvik on the west (see Chapter 4), a fact which construction has taken into account.

When complete, the Highway's new path will actually add some 35 miles, or 56km, to the total route length. However, it will eliminate the half-hour ferry which is frequently the cause of delays in summer when its capacity is over-taxed. About 200,000 pasengers annually use the Lyngseidet-Olderdalen ferry. This is unlikely to be withdrawn when the new road opens, for it will continue to serve the western side of the fjord and Route 91 (see Chapter 6).

In Nordland, the ferry across Leirfjorden between Sommerset and Bonnåsjøen takes only fifteen minutes, but demands for its replacement come from all quarters—not least the tourist authorities. The improvements and re-routeing of 1966, described in Chapter 3, do not satisfy the critics, who claim that a road such as the Arctic Highway should be without ferry links.

The problems at Leirfjorden are immense. There is no obvious bridging place at either Sommerset or Bonnåsjøen, or indeed between these points. To find a suitable site for a bridge would necessitate an extension of the Highway on both sides of the fjord towards the head. Even so, the costs of building a bridge would be very great. Now feasibility and comparative-cost studies have been directed to an alternative in the form of a road east of Sommerset to reach almost the head of the fjord, where a causeway or low-cost bridge may be built. The Highway would then strike north towards Mørsvikfjorden, cutting off about a dozen or so miles of the present road north of Bonnåsjøen.

The difficulties here are largely in the absence of an obvious route through the Sildhopfjell and Horndalfjellet. These mountains, capped by cirque glaciers, rise to over 3,000ft. Although the low ground around Kobbvatnet and the river Gjerdal, which discharge into the Leirfjord, makes much of this route attractive, the mountain divide poses a seemingly insoluble problem. At the time of writing, no definite decision has been made, but it is my own opinion that a bridge will be the eventual choice. In any case, it is unlikely that the Sommerset-Bonnåsjøen ferry will be superseded before the mid-1980s.

One of the original ferries seems destined to survive all change. This is the Tysfjord crossing between Bognes and Skarberget. The coast in this part of Nordland is so deeply indented that its fjords strètch back towards the Norwegian-Swedish border. Only the building of a series of perhaps some five or six costly bridges, coupled with a re-routeing of the Highway eastward to cross the inter-fjord peninsulas, could make this ferry unnecessary. No, it is surely here to stay and to provide drivers with a half-hour rest in which to enjoy the magnificent scenery of the coast. Pierhead delays will, no doubt, be reduced by the employment of additional ferry vessels.

The ferry across Kvænangsfjorden between Sørstraumen and Badderen was replaced by a road around the head of the fjord as long ago as the early 1940s, but a new bridge is now planned close to the old ferry crossing. Scheduled for 1974-7, the bridge will span the narrows which separate the outer from the inner fjord. This bottleneck will cut bridging costs and, when the construction is completed, the Highway will be shortened by about 20 miles, or 32km.

These, then, are some of the prospects for the future of the Arctic Highway, but it must be recalled that its rôle is that of a trunk road built to feed and be fed by its branches. The Highway is a vital part, but only a part, of the road system of arctic Norway, a system that is to undergo dramatic changes in the 1970s and 80s. The traffic using the Highway will greatly be affected by some of the transformations which the system will experience; the more pertinent changes are briefly summarised below. Reference should be made to Chapter 6, in which the existing branch roads are described.

Most of the major developments in the next two decades will be in Finnmark, where the road network is least dense. Here the plans envisage both improvements to the radial road system and the completion of an east-west interior road which would be complementary to the Highway.

Route 92 is due to be completed in the mid-1970s, when the Valjok-Lævvajok gap is closed. At roughly the same time the old snow-road between Karasjok and Kautokeino will be made into an all-weather vehicular road. West out of Kautokeino a new road already reaches Biddjovagge and the copper mines on the

Troms-Finnmark border. Eventually, probably in the 1980s, this will be extended to link with Route 865 at Bilto. When all these roads are built, they will form a complete interior route between the Tana bridge and Nordreisa—both on the Arctic Highway—almost linking those two great northern fjords: the Varangerfjord and Lyngenfjorden.

The two existing radial routes in Finnmark, Routes 93 and 96, are to be improved, the work on the former being nearly completed. When it is recalled that only a few years ago this road stopped short at Kautokeino and was little more than a track, the new re-routed international road with its largely oil-gravel surface is all the more remarkable and encouraging.

A new international radial road is planned to link the Highway with Finland. This is expected to be built in the present decade, for its Norwegian section will be only a few miles long, west out of Neiden. After crossing the Finnish border near Rajavartiosto, it will join the minor road and snowmobile route running along the complex of lakes that includes Inari, to finally join Route 4 at Kaamanen. When the Finnish road is reconstructed, the route will replace the pre-war international link between the old Finnish Arctic Road and Varangerfjorden which was lost in the Soviet annexation of the Petsamo Corridor.

A radial route which is to be extended north is Route 95. By the mid-1970s a new section will be built to link Repvåg with Kåfjord, but this is unlikely greatly to affect the Highway unless it increases tourist traffic to the North Cape.

Outside Finnmark, the two developments that are most certain to bring additional traffic on to the Arctic Highway are new international routes in Nordland linking Norway with Sweden. The earliest of these, due for completion in the mid-1970s, is in Junkerdalen (see Chapter 3). Here the minor valley road is to be re-built and extended beyond the Graddisfjellstue up to the border. From the frontier, a new Swedish highway will follow the line of existing tracks to Sädvaluspen, where it will join the road to that delightful Gulf of Bothnia town, Skellefteå.

A more ambitious project is planned for the 1980s. A road east from Narvik will cross into Sweden with the Ofot railway to Kirunavarra, where it will join the Swedish Route 98—again to the Gulf of Bothnia. If this plan matures, it will undoubtedly throw a heavy burden of traffic upon the Highway in the Narvik district. The railway already carries over 250,000 passengers across the frontier each year and a road would certainly be more attractive.

One final plan deserves at least a mention. This is the often-talked-of coast road, an alternative to the Highway. Few expect a continuous road ever to be built. It would be costly, incorporate too many ferries and bridges, and be of limited use except in its service to the scattered fishing settlements. What will almost certainly occur is the construction of links between existing coast roads and between the Highway and this system. Advanced planning is already under way and some sections are built. The areas most likely to be affected are those along the coast between the Ranafjord and Sørfolda and the already well-served coastal districts between Tromsø and Narvik. The possibility of a continuous coast road cannot entirely be dismissed, but that it should ever rival the Arctic Highway is unthinkable.

With such an exciting future in store for the road system of North Norway, the impact that these changes will have on the population and economy of the region will be no less far reaching. It is true that roads cannot cure the *mørkesyke*—the very real depression that results from long hours of winter darkness. It is true that only a very few will actually gain their livelihood directly from the road. New routes cannot change the environment; but an improved road system will allow the North Norwegians to cope more easily with the difficulties of their environment and exploit to the full their natural resources. Likewise, the Arctic Highway will encourage the growth of settlements along its length, though perhaps at the expense of the island communities. One particular *test case* of this tendency may be seen in the rival claims of Alta and Hammerfest to be the leading town of west Finnmark.

Hammerfest's economy is tied to fishing and the town has a

single land-link with the mainland. Alta's economy is capable of diversification—farming, fishing, forestry, quarrying and tourism, all are possible—and the town is ideally sited on the Arctic Highway at its junction with the international Route 93. Plans to build a new Alta City (the English word *city* is often used, much to the annoyance of some of the townspeople!) are already in hand. The enlarged Alta kommune has a town-planning office where there is talk of 'functional areas' and of a 'green invasion', the latter term being used to describe the inhabitants' growing awareness of the advantages of neatly ordered gardens and of the outward appearance of the town. When all the plans have been fulfilled, the population will have risen by 60 per cent, there will be 15,000 in the kommune, and Alta will be in an unassailable position as the far north's most important settlement. Much of the new *city* will be inland, south west of the Highway, on high ground between the fjord and Altælva. But a new herring port, employing 800, is planned for Kvænvik, on the Highway east of Alta, and the airport is to be expanded, possibly to include a 6,000ft runway. In the core of the new town there will be an extra 2,000 people; there will be a new school, a new hospital and an industrial estate where private companies will be invited to occupy the factory buildings provided by the kommune.

As these plans mature, the Highway and Route 93 will contribute to their success and derive benefits from the settlement's expansion. The through traffic will certainly grow at rates well above its present 5-10 per cent figure, but it is in the generation of traffic that the new Alta will affect the Highway most of all.

What is happening at Alta will be repeated to a greater or lesser degree in other Highway settlements. Mo i Rana, Narvik, Lakselv and Kirkenes spring particularly to mind. As it begins to play a more important rôle, the Highway will have much to contribute to North Norway's growth in the 1970s and 80s.

A Note on Tourism

Tourism in North Norway is relatively underdeveloped; its

rate of growth is well below what could be expected of an area which needs to diversify its economy and find new sources of income. Perhaps too much attention has been paid to the more tangible natural resources and too little investment has been made in invisibles. Yet when one considers the way in which countries such as Iceland, and even Greenland, have succeeded in attracting overseas visitors, the failure to realise the enormous potential which tourism has in arctic Norway is almost inexplicable. Certainly there is a lack of effective publicity, as a visit to any tourist agency will show. Even in the State-sponsored agencies there is an abysmal ignorance of North Norway on the part of those whose job it is to *sell* the region. A courteous reception is little compensation for a lack of information. Popular misconceptions—as to climate, hotels, transport and attractions—go uncorrected. This is not the place to suggest remedies, but tourism in the North probably is capable of greater expansion than any other single economic activity if tackled in a positive and vigorous manner.

In Norway as a whole, the 1960s showed a 113 per cent rise in tourist traffic, but North Norway's share failed to show a proportional increase. Indeed, in recent years figures based on hotel bookings show that if it were not for a slight increase in Norwegian visitors to Troms and Finnmark, there would have been an absolute decline in numbers. Figures for foreign tourists almost everywhere show a reduction. Despite construction of new hotels, such as that at Kautokeino (on Route 93), and the expansion of others, for example in Bodø and Lakselv, the total number of persons employed in hotels and restaurants in North Norway has risen only 15 per cent in the summer seasons between 1960 and 1970.

The Arctic Highway's contribution to the tourist industry has yet to come. In the past, the positive attractions of scenery, fishing and the midnight sun, as well as the real pleasures of isolation in this most unspoilt part of Europe, have been overshadowed by the problem of distance. Many tourists have found it easier and quicker to take the Coastal Express and view the country from deck or sun lounge, getting to know neither land nor people. Only those with time to spare have come by road from the south and even fewer have got beyond Narvik or Tromsø. Motoring organisations must take their share of the blame, for suggested daily journeys of 100 miles or so must deter those who might wish to travel the 1,800-mile round trip from Mo to Kirkenes— knowing that Oslo is another 660 miles south of Mo and that Bergen is 80 miles further. In fact, with careful driving, much higher daily averages are possible except on the worst sections in east Finnmark. In good weather, and with a co-driver, I have driven the 650 miles from west of Alta to Kemi, on the Gulf of Bothnia, in a day, using the Karasjok route. As improvements to the Highway are made and ferries reduced, daily milages must rise. Even in summer, fast driving may, however, require some special techniques or precautionary measures and the motorist new to the North is certainly wise to seek advice.

If the transport of cars to arctic Norway were easier or there was an expansion of car hire, then the time-consuming journey to the North could be cut. Likewise, there seems little reason why coach-touring, with charter air travel to the North, should not become a feature of the Highway in summer, for the North Norway Bus from Fauske is not really suited to carrying holiday-makers. One alternative, the travelling hotel, has already been tried (*see plate, page 176*).

Once on the Highway, the tourist will never wish to turn back. The scenery is such that no words can do justice to the wild beauty on which the eye may feast. This is a land where one may wander at will, without traffic jams or over-crowding, and where all the pressures of the twentieth century seem to be relaxed. Few who can afford to do otherwise visit North Norway only once. The annual visitor is commonplace.

The Highway is the obvious tourist route. Not only is there a great deal to see from the road but, by its branches, the whole region is accessible. Camping just off the road is possible, either at the growing number of simple but well-run official sites or on the tracts of open land through which it passes. As the Highway is raised above snow-drift levels, it becomes increasingly necessary to follow tracks off the road to find suitable sites but, in the light of the midnight sun and with only the sound of a bubbling waterfall to break the silence, this is camping at its best.

Hotels on the Highway—and I think I have visited them all—vary from the small *gjestgiveri* of villages such as Innhavet, to the larger hotels of towns like Narvik. If one expects to find a choice of bars and a floor show, then it is better to stay in Oslo or Bergen, but the North's hospitality, cleanliness and helpfulness are unequalled. Hotels with a wide range of amenities are to be found in Mo i Rana, Narvik and Kirkenes or, of course, off the Highway in towns such as Tromsø. In Alta, Lakselv/Banak and Bardufoss there are moderate sized establishments, and the excellent Polar Circle hotel is mentioned in Chapter 3. Some confusion arises over nomenclature: *gjestgiveri, hospits, turisthotell, høyfjellhotell*. In fact the names mislead, and experience is the best guide! Accommodation of the bed-and-breakfast kind is also available on the Highway at farms advertising on roadside boards in Norwegian or Finnish.

There is a real need for one or two new hotels on the Highway, sited where they would best serve the needs of the tourist, that is at major route junctions and preferably near an airport. The construction of a large hotel at Kautokeino (see Chapter 6) was, in my opinion, a mistake. Had the same building been erected ninety miles further north at Alta it would have served a more useful purpose.

New hotels are planned along the Highway at Bardufoss and in the expanded Alta, but much of the development is, or will be, away from this road. Naturally enough, Bodø and Tromsø will have the lion's share of new accommodation with, respectively, the SAS Royal Hotel and the Royal Hotel (formerly the Central). However, the new buildings planned for Vardø and on Magerøya might seem ill-sited or, at least, inadequate compared with alternative positions at, say, Rustefjelbma or Varangerbotn and Russenes. The Målselv district and Nordkjosbotn are other places along the Highway where more accommodation could prove

P

profitable if road touring increases. Even on the limited information available, the present approximate 15 per cent annual increase in road touring on the Highway by foreign visitors should encourage further hotel building. Especially noteworthy is the recent trend of airlines to become directly interested in hotel projects in North Norway. Perhaps the air-travel and coach-touring holiday, already referred to, is not so far off. The five-fold increase in accommodation in the last twenty-five years in North Norway may well be exceeded in the next quarter of a century.

To hitchhike through North Norway on the Arctic Highway may be too adventurous for most, but is not unknown. It might be noted, however, that local people usually follow that old-fashioned custom of offering payment for lifts received!

Many tourists are attracted by the unrivalled opportunities to fish in North Norway and the Highway crosses some of the best salmon rivers in the world. Finnmark is an angler's paradise. Although salmon fishing on rivers like the Alta, Tana, Laks or Neiden is expensive, one does not have to be an English duke or American millionaire to fish the innumerable lakes and smaller rivers seemingly full of trout. Sea fishing, of course, is also possible.

Few visitors to the North will go to fly, yet armed with a pilot's licence—preferably with a seaplane rating—there can be no better way of seeing the country. At least two flying clubs, at Alta and at Mo i Rana, will loan aeroplanes at charges which are better than reasonable and with a lack of formality which is as commendable as it is surprising. In a seaplane it is possible to penetrate the interior and land on lake or river. It is the ideal transport for a fishing holiday combined with road touring.

It is arguable which is the best time of year to travel the Highway. The climate is undoubtedly best in summer, but the mosquito is at its most bellicose. Autumn is the most colourful time of the year, but the frosts of the *jernatter* are less attractive. Spring is a special season in all arctic regions, when the sun returns and the snows clear to reveal flower carpets. It is traditionally a festive season, but it is a poor period for driving, especially on gravel roads. The brave may drive in winter, if they are prepared to limit milages, find roads closed and familiarise themselves with the techniques of driving on snow that are second nature to the local people. Books such as Iva Maasing's *Vinterkørsel* provide a useful preparation.

For a variety of reasons summer is, in fact, the tourist season, with July as the most popular month. Certainly this allows the visitor to make the most of his holiday in the long hours of daylight, when it never seems dark enough to retire for the night.

If the Highway did no more than introduce just a few visitors each year to this arctic wonderland, its construction would have been justified.

Appendix A

Road Costs

Nordland (whole fylke)

Year	Improvement	Maintenance
1964	23.7	6.9
1965	27.2	7.4
1966	23.2	8.9
1967	22.5	16.3
1968	28.8	12.8
1969	28.2	18.4

Troms

Year	Improvement	Maintenance
1964	4.3	5.1
1965	4.3	5.3
1966	4.2	5.5
1967	5.0	5.9
1968	4.8	6.0
1969	5.1	6.2

Finnmark

Year	Improvement	Maintenance
1964	3.3	4.5
1965	3.7	5.2
1966	3.2	6.3
1967	6.8	7.2
1968	7.2	7.4
1969	7.7	7.6

Table III Annual cost of maintaining and improving all roads in Arctic Norway, in millions of Norwegian kroner.

	All Roads	Arctic Highway
Nordland (whole fylke)	186.0	81.6
Troms	86.9	30.7
Finnmark	204.0	29.3

Table IV Per capita expenditure on all roads and on the Arctic Highway in 1969, in Norwegian kroner.

	1964	1965	1966	1967	1968	1969
Nordland (whole fylke)	17.3	17.8	17.0	16.5	18.3	20.4
Troms	2.7	2.7	2.7	2.6	3.5	4.0
Finnmark	1.6	1.6	1.6	1.0	3.0	2.2

Table V Annual expenditure on construction and maintenance of Arctic Highway (E/R6 south of Mo i Rana), in millions of Norwegian kroner.

Note 1. By reference to Appendix C, Table X, further comparisons between the fylker may be made.
2. Source of raw statistics on costs: *Statens Vegvesen, Vegdirektoratet, Oslo.*

Appendix B

Traffic Census Figures

Note: A full-scale traffic census is held in North Norway once every five years, with sporadic checks in the inter-censal years. Although the counting is generally not automatic and the periods during which the count is taken are short, it yields some useful information. The last full census for which figures are available was in 1965.

Census Point	Annual Average vpd	Summer Average vpd
Mo i Rana	1,546	1,974
Lønsdal[1]	-	409
Fauske	417	606
Ulsvåg	402	507
Fossbakken	531	676
Brandvoll	776	911
Andselv	1,670	1,950
Heia	342	469
Vollan	557	764
Tretten[2]	210	305
Langfjordbotn	90	140
Skaidi	410	715
Lakselv	348	456
Ifjord[3]	-	99
Tanabru	279	389
Varangerbotn	362	446

Table VI   Vehicles per day at selected census points on the Arctic Highway in 1965. Annual average figures.

Notes 1. Polar Circle road. Closed in winter at time of census.
   2. North of Nordreisa.
   3. Easterly route closed in winter.

Source: Statens Vegvesen, Vegdirektoratet, Oslo.

Census Point	1960	1963	1965	1966	1967
Andselv	741	-	1,670	1,710	1,960
Vollan	435	-	557	795	880
Langfjordbotn	-	81	90	-	175
Varangerbotn	-	281	362	-	416

Table VII   Increases in traffic along the Arctic Highway in the 1960s, as exemplified by four representative census points. Figures are the annual average number of vehicles per day.

Sources: *Statens Vegvesen, Vegdirektoratet, Oslo; Norges Automobil-Forbund, Oslo.*

Section of road		Section of road	
i.	*Mo i Rana town district*	ix.	Andenes town district
ii.	*Vensmoen - Saltdal*	x.	*Fossen - Olsborg*
iii.	Fauske - Bodø	xi.	Finnsnes town district
iv.	*Fauske - Vargåsen*	xii.	Vollan - Tromsø and Storsteines
v.	*Grindjord - Narvik*	xiii.	*Alta town district*
vi.	*Narvik - Rombakfjord*	xiv.	*Skallenes - Banak*
vii.	Harstad - Gare	xv.	Vadsø - Jakobselv
viii.	Harstad - Sandtorg	xvi.	*Neiden - Kirkenes*

Table VIII   High traffic densities on roads in arctic Norway. Sections of main extra-urban highway which exceeded annual average of 500 vehicles per day in 1965. Those sections shown in *italics* are, at least in part, the Arctic Highway.

Source: *Trafikkart over Norge 1965.*

Appendix C

The Tortuosity of the Arctic Highway

Frequent reference is made in the text (see especially Chapter 1) to the difficulties of constructing a direct route between any two places in North Norway. Various methods for quantifying this problem can be used, such as the *tortuosity ratio* or the *detour index*. The latter procedure is applied to the Arctic Highway in the table below.

	The route between:	Arctic Highway Detour Index
Nordland:	Mo i Rana and the Polar Circle	140
	The Polar Circle and Rognan	116
	Rognan and Fauske	165
	Fauske and Sommerset	139
	Bonnåsjøen and Bognes	143
	Skarberget and Narvik	153
	Narvik and Bjerkvik	258
Troms :	Bjerkvik and Fossbakken	130
	Fossbakken and Bardu	123
	Bardu and Nordkjosbotn	151
	Nordkjosbotn and Lyngseidet	123
	Olderdalen and Nordreisa	185
	Nordreisa and Kvænangsbotn	181
	Kvænangsbotn and Burfjord	140
Finnmark:	Burfjord and Alta	217
	Alta and Skaidi	118
	Skaidi and Olderfjord	107
	Olderfjord and Lakselv	133
	Lakselv and Børselv	110

Børselv and Ifjord	131
Ifjord and Varangerbotn	167
Varangerbotn and Kirkenes	170

Table IX The tortuosity of the Arctic Highway as shown by its detour indices.

The detour index is given by the simple formula:

$$DI = \frac{R}{D} \times 100$$

where DI is the detour index; R is the distance by road and D is the direct distance. Hence, indices close to 100 indicate a high degree of directness. An index of 200 indicates that the distance by road is twice that *as the crow flies*. Because calculations are based upon an assumed function for the Highway, the index should be taken only as a general indication of directness.

Fylke Comparisons

The table below may be used, in conjunction with the statistics in *Appendix A*, to give further comparisons between the fylker in terms of expenditure on the Highway.

Fylke		Area in square miles	Length of Arctic Highway in miles
Nordland:	whole fylke	14,728	-
	north of Mo i Rana	10,000	287
Troms		10,005	240
Finnmark		18,580	381

Table X Fylke comparisons. The length of Highway within, and the area of each fylke.

Appendix D

Distances along the Highway

All distances are approximations to the nearest whole mile or kilometre. Re-alignments and major improvements to the Highway may call for adjustments to the table below. All major branch junctions with the Highway are included; for distances along branch roads see Chapter 6.

Towards Kirkenes			Towards Mo i Rana	
km	miles		miles	km
0	0	Mo i Rana	908	1,462
12	7	Røsvoll	901	1,450
80	49	Polar Circle	859	1,382
152	94	Rognan	814	1,310
183	113	Fauske	795	1,279
191	118	Vargåsen	790	1,271
231	143	Sommerset	765	1,231
		(ferry: approx 15min)		
231	143	Bonnåsjøen	765	1,231
290	180	Innhavet	728	1,172
317	197	Ulsvåg	711	1,145
338	209	Bognes	699	1,124
		(ferry: approx 30min)		
338	209	Skarberget	699	1,124
353	219	Sætran	689	1,109
361	224	Forså	684	1,101
378	235	Ballangen	673	1,084
402	249	Grindjord	659	1,060
421	261	Narvik	647	1,041
456	283	Bjerkvik	625	1,006
487	302	Fossbakken	606	975
509	316	Brandvoll	592	953
518	321	Bardu	587	944
537	333	Elverum	575	925

Towards Kirkenes			Towards Mo i Rana	
km	miles		miles	km
545	338	Andselv/Bardufoss	570	917
555	344	Buktamo	564	907
559	347	Olsborg	561	903
580	360	Heia	548	882
596	370	Storsteinnes	538	866
613	381	Nordkjosbotn	527	849
622	386	Øvergård	522	840
631	392	Oteren	516	831
673	418	Lyngseidet	490	789
		(ferry: approx 30min)		
673	418	Olderdalen	490	789
707	439	Brinken	469	755
718	446	Sørkjosen	462	744
723	449	Nordreisa	459	739
776	482	Karvik	426	686
799	496	Kvænangsbotn	412	663
833	517	Burfjord	391	629
917	570	Kåfjord	338	545
934	580	Alta	328	528
1,021	634	Skaidi	274	441
1,044	648	Olderfjord	260	418
1,108	688	Lakselv	220	354
1,150	714	Børselv	194	312
1,231	764	Ifjord	144	231
1,296	805	Rustefjelbma	103	166
1,319	819	Tana bru	89	143
1,337	830	Varangerbotn	78	125
1,344	835	Karlebotn	73	118
1,418	881	Neiden	27	44
1,456	904	Hesseng	4	6
1,462	908	Kirkenes	0	0

Appendix E

Glossary

The words in this short glossary occur in the text or appear on maps of North Norway. It should be noted that there are two forms of Norwegian in use (Bokmål and Nynorsk) and that, especially in Finnmark, the use of Lappish and even Finnish gives alternative place names. Further confusion occurs through repeated use of the same place names in different locations.

The definite article is given to a word by a suffix (*-a, -en* or *-et*). Plural forms are generally the suffix *-er* or *-ene*. Many of the words below appear as the suffix to a place name.

amt	*county* (until 1918)
bakke	hill
bane	railway
berg	rock
botn	fjord head or cirque
bre	glacier
bru	bridge
bukt	bay
by	town
dal	valley
eid	isthmus, col
elv	river
fisk	fish
fjell	mountain divide
flat	plain
flyplass	airfield

foss	waterfall
fylke	*county*
gate	street
gjestgiveri	guest-house
gård	house, farm
haug	hill
havn	harbour
holme	small island
indre	inner
is	ice or glacier
jokka	river
jord	ground, farmland
kai	quay
kil	creek
kirke	church
kjos	small inlet
kollen	knoll
kommune	district (part of a *county*)
laks	salmon
lille	little
myr	marsh
nedre	lower
nord	north
skjær	skerry
skog	wood
sogn	parish
sted	place
stein	stone
stor	big
sør	south
tind	mountain peak
tjern	small lake
topp	mountain summit
vann	lake
vatn	lake
vei	road

vest	west
vidde	plateau
vik	creek, inlet
vinter	winter
voll	meadow
øst	east
øy	island
ås	hill

Acknowledgements

I am greatly indebted to innumerable friends and officials in Norway who have assisted me in the preparation of this book. In particular I wish to thank Herr Johs. Hedemann of Statens Vegvesen, Vegdirektoratet in Oslo and the Highway Authorities at Bodø, Tromsø and Vadsø, who have been so willing to spend time in answering my questions, to grant me interviews and to provide me with maps and statistics. The assistance was theirs, but any errors of fact or interpretation are mine.

My travels in North Norway have been made all the more enjoyable and interesting through the unfailing kindness of both the ordinary people and those in official positions. Especially I wish to acknowledge the help given by senior officials in Tromsø and Alta, by Fru Aase Wright of Mo i Rana, by the tourist authorities and hotel managements, and by the flying clubs of Alta and Mo i Rana.

Among others who have provided information or assistance are Herr Petter Hamnes, Traffic Manager of Nord-Norge-Busen; Herr Mikal Bjørklid of Bjørklids Ferjerederi; Herr Olav Lande of Kirkenes; Herr J. Kraft Johanssen, Managing Director of Aktieselskabet Sydvaranger; Major Archie Douglas, MBE, late Grenadier Guards; the Scandinavian Airline System and the Norges Automobil-Forbund. To them I express my gratitude.

My work with the Lapps of Finnmark was made possible by Ellen Anne Hatta of Masi, herself a Lapp, who acted as guide and interpreter. To her I owe a special debt, for it was she who introduced me to the real Lapland.

Assistance with original research and the compilation of statistics has been given by a number of my students and I am grateful for their help.

The map of Aisaroiui was first published in *Nomads of the Arctic,* my account of the Kautokeino Lapps. The reproduction on page 137 is from *The Geographical Magazine,* London, by kind permission.

Finally, I wish to thank my father and Dr J. B. Harley for their general advice on and help with the preparation of the book.

JOHN DOUGLAS
London, Easter 1971

Index

Notes: 1   *It should be noted that the letters æ, ø and å come at the end of the alphabet.*

2   *In general, valleys are indexed under the name of the river, thus : Målselvdalen is listed under Målselva. Places taking their name from the river on which they are sited are separately indexed, thus : Lakselv (village); Lakselva (river).*

3   *Where the same place-name is shared by more than one location mentioned in the text, the Kommune or District is given in parentheses.*

4   *Plate references appear in* **bold type.**

Q